# MATH ANXIETY:
## What It Is and
## What To Do About It

Charlie Mitchell, Ph.D.
Lauren Collins, M.Ed.

C. 1

**KENDALL/HUNT PUBLISHING COMPANY**
2460 Kerper Boulevard  P.O. Box 539  Dubuque, Iowa 52004-0539

*For Andy and Matthew*
*with love*

# CONTENTS

# ACKNOWLEDGMENTS

This book was inspired by the many people with math anxiety who have come to me for counseling over the years. There is nothing miraculous about their "cure," yet their gratitude for helping them overcome the problem has been overwhelming. Such appreciation has motivated me to share my approach with others.

My intent is not to introduce new concepts or practices in counseling psychology to those professionals engaged in its practice. Rather, this book takes well-known, psychotherapeutic strategies and applies them to the treatment of the specific problem of math anxiety. The professional will recognize major contributions of theory and practice by the following people: Albert Ellis, Rational Emotive Therapy; Joseph Wolpe, Systematic Desensitization; Edmund Jacobson, Progressive Relaxation; John Grinder and Richard Bandler, Neuro-Linguistic Programming; Donald Meichenbaum, Stress Innoculation and Cognitive Behavior Modification; Richard Suinn and F. Richardson, Anxiety Management Training; and M. Goldfield, Systematic Desensitization and Cognitive Change Methods.

Two books written specifically on the subject of math anxiety which deal with the mathematical aspects of the problem must also be acknowledged here as significant contributing works: *Mind Over Math* by Stanley Kogelman and Joseph Warren, and *Overcoming Math Anxiety* by Sheila Tobias. Both of these titles are very helpful in conjunction with a psychological treatment of math anxiety.

The work of other people in the areas of stress and anxiety management should also be acknowledged. Any attempt, however, to list all contributors would in itself be book length. With that in mind, I would like to recognize the contributions of the following researchers and practitioners whose work is directly related to the description and management of math anxiety (the titles of their work can be found in Appendix II): G.J. Allen, B. Baker, A. Bandura, A. Beck, H.H. Dawsley, J.L. Deffenbacher, D.R. Denny and P.A. Rupert, L. Donner, I. Farber, D.E. Friedman, M. Kahn, P. Lang, R. Liebert, A.A. Lazarus, M. Mahoney, L. Morris, R.A. Osterhouse, G. Paul, S. Schacter, J.F. Sipich, B.F. Skinner, M. Spiegler, C. Spielberger, and R. Zemore.

Charlie Mitchell, Ph.D.

# FOREWORD

Dr. Mitchell has succeeded in writing a highly readable and practical book for individuals experiencing math anxiety. His book is a must for anyone who becomes nervous when dealing with numbers. Dr. Mitchell seems to take the reader gently by the hand, guiding the individual through the maze of irrational thoughts and fostering skill development to actively counteract anxiety.

Readers will easily be able to identify with the many lucid examples of physiological and psychological reactions to stress that are presented throughout this well-written book. A superior self-paced approach will draw out the arm chair psychologist in all readers to determine the roots of one's math avoidance and anxiety. Dr. Mitchell also explodes math myths, those burdens that individuals carry which contribute to one's sense of incompetence and defeatism.

Self-growth and self-understanding await the reader, a challenge that is well worth the effort.

Mamiko Odegard, Ph.D.
Clinical Psychologist

# INTRODUCTION

For many individuals, math anxiety is a real and disabling problem. This book was written to offer relief from that problem — for those who suffer from math anxiety; for school counselors who work with math-anxious students; and for math teachers who want to understand the dynamics of what some of their students are experiencing.

*Math Anxiety: What It Is and What To Do About It* offers both a description of and a solution for math anxiety. It approaches the causes and the remedies from three interrelated perspectives: the physical symptoms of the problem; the cognitive images and beliefs which create the problem; and avoidance conditioning which perpetuates the problem.

When math anxiety disables a person, several things happen simultaneously. The body, reacting as if it is being threatened, releases chemicals in response to the danger; the mind carries on an internal monologue; and the autonomic nervous system automatically reacts by mobilizing the body's physical resources for self-protection. A cycle of anxiety is put into motion as the body, mind, and emotions reinforce each other. Avoidance of the threat, math, is likely to follow.

In order to understand math anxiety, these three aspects must be examined separately as well as how they interact. The same is true for the treatment of math anxiety. In order to break the anxiety cycle, intervention strategies should involve the body, the mind, and the autonomic nervous system. Each aspect is a contributing factor that must be transformed to work positively rather than against a person.

Accordingly, the book addresses, chapter by chapter, each cause of math anxiety, then offers solutions for that aspect of the problem. The intervention strategies described are systematic relaxation, which helps to control the disabling physical symptoms of anxiety; cognitive restructuring, which undoes the self-defeating thinking processes that cause math anxiety; and desensitization, which eliminates the conditioned or reflexive behaviors that are disabling.

Most people can reduce or eliminate their math anxiety on their own, by following the exercises prescribed in this book. The first step toward resolution is understanding what math anxiety is and how math anxiety occurs. Understanding is, however, only a first step. In order to "undo" the existing pattern which perpetuates the anxiety, some "re-programming" activities must also take place. The reader must actively perform the recommended exercises contained in this book to produce results. In some serious cases, more extensive treatment by a psychologist, counselor, or other professional is recommended.

# Chapter 1

# ARE YOU A VICTIM
# OF MATH ANXIETY?

Have you ever prepared for a math test and felt sure the night before the test that you knew the material? The next day you go to class to take the test and as you arrive at the classroom you have a vague feeling of nauseousness. As you enter the classroom you are aware that your heart is beating faster and that your muscles are tense. By the time the test is handed out you are perspiring, your hands and feet are cold and clammy, your heart is pounding, and with each heartbeat your head as well as your chest seem to pulsate. You feel as if you are going to explode with the excess energy that is careening through your system. You feel as if you would like to spring from your desk and sprint from the classroom. As you gain a measure of control over your impulse to flee you are aware that you have lost the capacity to recall what you studied the night before. You do not recognize math signs, symbols, or processes. You stare at the math test as if it were completely foreign to you. You experience a "mental block." Your mind refuses to work and you feel frustrated, upset, and humiliated.

After you leave the classroom an amazing thing happens: you begin to remember bits and pieces of what was on the test. Within half an hour you can recall not only most of what was on the test but also how to work the problems.

If you have experienced this situation, you have experienced **MATH ANXIETY**.

If you have ever been handed the check for a group of people having lunch together, asked to figure how much each person owes including tips, and out of embarrassment decided to pay the entire bill rather than attempt the calculations, you were probably experiencing **MATH ANXIETY**.

If you hate to play card games that require that you think mathematically, chances are good that you have **MATH ANXIETY**.

If you live in fear of having to interpret graphs, job trends, statistics; to do bookkeeping or income tax reports; to figure mark-ups, mark-downs, percentages, proportions, rates, or interest on loans; it is likely that you have **MATH ANXIETY**.

If you detest balancing your checkbook — and therefore never do it — you may have **MATH ANXIETY**.

If you chose your career because there was no math involved or because you did not have to take math in college to get into your career field, chances are very good that you have **MATH ANXIETY**.

If you experience **all** of these symptoms, you do indeed have **MATH ANXIETY**, and it is you for whom this book is written.

# Chapter 2

# THE PHYSICAL ASPECTS
# OF MATH ANXIETY

Math anxiety is caused by a combination of physical, cognitive, and psychobehavioral components. Although they all work together to create the problem of math anxiety, each component needs to be examined separately.

The physical aspects of math anxiety are biological, consisting of hormonal, chemical, and muscular changes in the body. The by-product of this biological condition is a disability in thinking.

The physical symptoms of math anxiety can be very intense for some people and almost negligible for others. Those who experience intense anxiety are usually aware of its presence and know that they are cognitively disabled by it. On the other hand, some people have little or no awareness of the presence of anxiety and its effects. Nevertheless, their diminished ability to perform intellectually will unmistakenly indicate that anxiety is present.

The physical manifestations of anxiety include: muscle tension, "butterflies," nausea, shortness of breath, constriction in the throat, perspiration, clammy hands and feet, feeling faint, excess energy, tunnel vision, rapid heartbeat, increased blood pressure, and heightened sensory awareness. **When these symptoms are present, one's ability to think is greatly diminished**. Basic thinking processes like remembering, analyzing, synthesizing, and making generalizations from data, are affected by the presence of anxiety. The greater the intensity of the physical reaction to anxiety, the greater one's thinking processes will be disabled.

> **THE DEGREE TO WHICH A PERSON EXPERIENCES THE PHYSICAL AND CHEMICAL CHANGES IN THE ANXIETY RESPONSE IS THE DEGREE TO WHICH HIS OR HER ABILITY TO THINK IS DISABLED**

The biological state of fear or anxiety occurs in response to a perceived threat. The purpose of this biological reaction is to keep the body vigilant, ready to run or fight if necessary. This is called the "vigilance" or the "fight-or-flight" response. It is an attempt on the part of the body to insure self-protection by mobilizing its resources to that end.

The fight-or-flight response is a positive force that can be lifesaving. Historically, running or fighting has been an effective way of coping with threat. However, a problem occurs in present-day situations when the thing feared cannot be run from or fought. Examples of such fears are fear of failure, fear of being laughed at, fear of appearing foolish or stupid, and fear of not achieving one's life goals. These fears which cannot be run from or fought are called anxieties and are the sabre-toothed tigers of a modern society. It is fears such as looking stupid, experiencing humiliation, and losing self-esteem which underlie math anxiety.

A description of the biological aspects of anxiety is provided in the next paragraph. The purpose is to emphasize that the processes which disable a person mentally are explainable and not the result of magic. They are definable, physical processes which have the purpose of protecting the individual from harm by mobilizing physical resources.

Neurons on the outer edge of the brain send messages to the hypothalamus. The hypothalamus releases chemical "messengers" that activate the pituitary gland. The pituitary secretes a hormone which eventually activates the outer layer of the adrenal glands. This speeds up the body metabolism and changes blood sugars to a highly usable form. On another level, similar hormonal "messengers" trigger electro-chemical impulses that move down the brain stem, through the spinal cord, and into the core of the adrenal glands which release adrenaline. Adrenaline helps supply extra glucose to serve as fuel for the muscles, speeds up the heartbeat, and raises blood pressure. As a result of this process the blood vessels retract from the surface of the skin. As the blood vessels become smaller, blood pressure is increased and less blood is actively traveling through the major vessels. The excess blood pools in the capillaries around vital organs. The skin's surface temperature decreases. Perspiration further contributes to the lowering of the skin's temperature. One's attention is on the "outside" of the body scanning the environment looking for possible sources of threat. Muscles are tensed, ready to respond immediately if there is a need to do so. The body is in a state of alert, ready for a potential emergency.

A chart is provided on page 19 to show the correspondence between the physical symptoms of math anxiety and the internal biological process.

In summary, in a threatening situation the body mobilizes its resources for the purpose of self-protection. The degree to which a person experiences this physical condition is the degree to which that person's complex thinking processes of recognition, recall, analysis, synthesis, and generalization are disabled. This state is often called a mental block.

The next time you are unable to function when you are required to figure a restaurant bill, or are confronted with a math problem, or required to take a math test, take a minute and reflect on your physical state. The chances are great that you will find yourself experiencing some or all of the symptoms described earlier: muscle tension, "butterflies," nausea, shortness of breath, constriction in the throat, perspiration, clammy hands and feet, feeling faint, excess energy, tunnel vision, rapid heartbeat, increased blood pressure, and heightened awareness of surroundings, as well as an inability to perform complex intellectual tasks.

Let these symptoms be cues for you that something is going on inside you which will inhibit your ability to think. From this point of view, math anxiety is the presence of the physical state of "vigilance" or "fight-or-flight" which inhibits the thinking processes to the point where it is impossible to think mathematically or to perform mathematical computations.

However, you need not be the passive victim of this condition. You can do something about it.

| PHYSICAL SYMPTOM | BIOLOGICAL PROCESS |
|---|---|
| Butterflies, Nausea | Blood pooling in the capillaries around vital organs |
| Feeling faint | Increased blood pressure, decreased blood vessel size, blood "draining" from the head |
| Pulsating heart | Increased blood pressure |
| Clamminess | Blood vessels receding from the surface of the skin, combined with increased perspiration |
| Shortness of breath | Muscle tension in the diaphragm, increased need for oxygen |
| Shakiness | Tired, flexed muscles |
| Hot flashes | Increased blood supply to vital organs |
| Excess energy | Blood sugars from the liver, insulin from the pancreas |
| Muscle tension | Readiness to run or fight |
| Achiness | Lactic acid stored in the muscles |
| Perspiration | Cooling of the body temperature to make it more efficient for running or fighting |

## Exercise 2-A

Describe the physical effects of anxiety that you have experienced in a math-related situation (such as during a math test) by marking the list of symptoms below.

Check all that apply:

_____ muscle tension
_____ achiness
_____ stomach butterflies
_____ nausea
_____ feeling faint
_____ shakiness
_____ perspiration
_____ clammy hands
_____ clammy feet
_____ shortness of breath
_____ constriction in the throat
_____ rapid heartbeat
_____ pounding head
_____ increased blood pressure
_____ tunnel vision
_____ heightened sensory awareness
_____ excess energy
_____ hot flashes
_____ mental block

Refer to the chart on page 19 for an explanation of the biological processes behind the physical symptoms.

# Chapter 3

# SYSTEMATIC RELAXATION

The most effective place to begin an attack on math anxiety is at the physical symptom level because it is the physical state that disables the thinking processes. Relaxation is the technique to use to control or reverse the physical effects of math anxiety. Even people who show few obvious signs of physical anxiety, need to learn to use the relaxation technique.

Every person reading this book has the capacity to relax his or her muscles at will. With the advent of biofeedback, scientists have proven something that mystics and gurus have known for centuries. Scientists have discovered that humans have the capacity to relax their muscles, regulate their blood pressure, regulate their body temperature, etc. — at will. Biofeedback has given people undeniable evidence that they can control their physiological state. Armed with this knowledge people should no longer view themselves as passive victims of overwhelming anxiety.

Instead, they can choose the amount of tension they are willing to experience.

To demonstrate that you have control of your muscles, just reach down and place this book on the floor in front of you. In order for you to do that you must flex some muscles and relax others. Your muscles are controlled by your thoughts. If you think "arm, move" you will discover that your arm will move. You think and then respond. If you close your eyes and think about relaxing your stomach muscles you will discover that they will indeed relax. There is no magic involved. If you want to move your arm, you just move your arm. If you want to relax your stomach you just relax your stomach...or neck or back or shoulders. You are in control of your muscular state. You must keep just one thing in mind: there is nothing supernatural involved in relaxing. You simply think "relax," and your muscles will relax.

The reader might ask at this point, "so what?" So what if people have the capacity to relax at will? The point to be made here is that if people have the capacity to relax at will, they have the capacity to relieve themselves of the very symptoms that disable their ability to think clearly. They have the power to "cure" themselves of math anxiety.

Relaxation is a state that we each create within ourselves. You can call it other names like yoga, mind control, hypnosis, meditation, etc., or you can call it relaxation. Relaxation is a demonstration that people control their own internal environment. Because people have the capacity to relax at will, they have the capacity to relieve themselves of the very symptoms that disable their ability to think clearly. They have the power to "cure" themselves of the disabling symptoms of math anxiety.

The goal of the first intervention strategy, systematic relaxation, is to help you learn to relax muscles by tensing certain muscle groups and then relaxing them. Note that you cannot be both anxious and relaxed at the same time; the two states are mutually exclusive. Systematic relaxation will enable you to recognize the difference between tension and relaxation so that your awareness of muscle tension can serve as a cue that you need to relax. After learning to relax you will be able to induce relaxation at the first signs of tension.

There are many ways to relax, but here is one systematic procedure you can follow step by step. You might want to make a tape recording of the following instructions and then play them back and perform each step as you listen.

## The Relaxation Procedure

*Move your arms toward the center of your body and bend both arms at the elbow. Tighten your hands into fists and simultaneously tense the muscles in your upper arms and shoulders. Hold for ten seconds and then relax for fifteen to twenty seconds. Repeat this exercise one or two more times.*

*Tense your face muscles by wrinkling your forehead and pursing your lips. Hold for ten seconds. Then relax for fifteen to twenty seconds. Repeat this exercise one or two more times.*

*Take a deep breath and push out your stomach for eight to ten seconds. Then exhale and relax for fifteen to twenty seconds. Repeat this exercise one or two more times.*

*With your legs supported on a footrest, straighten both legs. Tense the muscles of your entire leg and pull your toes toward your head (keeping your feet on the stool). Hold for eight to ten seconds and then relax for fifteen to twenty seconds. Repeat this exercise one or two more times.*

*Close your eyes and turn off the world around you by focusing on what is taking place within your body.*

*Become aware of your heart beating.*

*Concentrate on slowing your heartbeat.*

*Become aware of the depth of your breathing and whether you are breathing deeply enough to relax.*

*Take a couple of deep breaths.*

*Try to breathe in relaxation and exhale tension.*

*Become aware of the tension in your system and where the tension is located.*

*As you become aware of the location of the tension, concentrate on relaxing those muscles. Perhaps the tension is located in your stomach muscles, your lower back, your neck, across the tops of your shoulders, in your forehead, or your jaw. Just concentrate on relaxing the muscles in your body that are tense.*

*As you concentrate on relaxing these muscles you will discover that they will relax. You are in control of your muscles and you will feel them relax.*

*Just let go and relax.*

*Now as you breathe slowly and deeply, concentrate on relaxing your stomach muscles.*

*Relax your diaphragm so you can breathe slowly and deeply.*

*Relax your scalp. Picture your scalp relaxing, and you will discover that your scalp is relaxing.*

*Relax your forehead, picture your forehead relaxing.*

*Picture your eyelids melting, melting.*

*Your eyes sinking, sinking.*

*Picture your face becoming very smooth.*

*Relax your jaw. The jaw is an area that often houses a great deal of tension. When you are tense, there is a tendency to clasp the teeth together tightly causing a point of pressure in the jaw.*

*So just let your jaw sag as you picture your face smooth, eyes sinking, sinking, eyelids melting, melting, forehead smooth, scalp relaxed, stomach muscles relaxed, as you breathe slowly and deeply.*

*Relax your neck. Relax your neck from the base of your skull downward.*

*Extend the relaxation downward through your shoulders. When you are tense, you have a tendency to lift your shoulders, causing the muscles across the tops of the shoulders to ache.*

*Just allow your shoulders to slump, as you relax deeply, completely.*

*Relax the big muscles in the middle of your back. Relax them from the tops downward.*

*As you relax the big muscles in your back, experience a calm, peaceful, serenity sweeping over your system, as you relax more and more deeply.*

*Relax the muscles in the small of your back.*

*Concentrate on letting go.*

*Relax the muscles in your hips, your thighs, and your calves.*

*Relax your feet and your toes.*

*Visualize any remaining tension in your system draining from the ends of your toes as you relax deeply and completely.*

*Your toes are relaxed, your calves relaxed, your thighs and hips relaxed, the small of your back relaxed, the big muscles in your back relaxed, your shoulders slumping, your arms, hands and fingers relaxed, neck muscles relaxed, jaw relaxed, face smooth, eyes sinking, sinking, eyelids melting, melting, forehead smooth, scalp relaxed, stomach muscles relaxed as you breathe slowly, deeply, and effectively.*

*Now visualize yourself in a scene in which you feel calm, peaceful, and serene. For example, you might picture yourself by the ocean with the waves rolling up onto the beach. You might visualize yourself beside a mountain stream, in a meadow, or in your favorite chair at home. But visualize your scene as vividly as you can and experience the peacefulness of this scene in the present.*

*Allow yourself to become "one" with the scene as you experience the tranquility of it.*

*Visualize the colors in your scene as clearly as you can and allow yourself to feel stimulated by the colors.*

*Listen to the sounds in your scene and experience the tranquility of the sounds.*

*Smell the odors in your scene and allow yourself to feel refreshed.*

*Experience the way your scene feels on your skin and feel bathed in renewal.*

*As you feel refreshed, renewed, tranquil, serene, notice that you feel centered, that is, at peace with yourself and at peace with the world around you.*

*As you are feeling centered, visualize yourself as you would like to be.*

*Feel comforted as you realize that the person you are visualizing is also you, a very large part of you, and that you are a positive person.*

*What a very pleasant realization indeed.*

*As you accept the positive part of yourself allow yourself to feel confidence and self-esteem.*

*Savor the very positive feelings it produces.*

*In this state of self-acceptance and relaxation, nurture and care for yourself. Feel as if you are your very best friend.*

*Now before you open your eyes, take just a moment to make a conscious assessment of just how very relaxed you really are.*

*Realize that you have produced this state of relaxation.*

*First you closed your eyes and turned off the world around you by focusing on what was taking place on the inside of your body. Secondly, you took several deep breaths, concentrating on inhaling relaxation and exhaling tension. Thirdly, you systematically relaxed your muscles throughout your body. Finally, you visualized a scene in which you could relax thoroughly.*

*Open your eyes as you continue to feel calm and refreshed. Be clear at this point about who created your relaxed state:* **you** *created the relaxed state. It did not come from somewhere outside you. It was the direct by-product of your thinking and the images you created. The same is also true, of course, with tension. Tension is the result of negative thinking and negative images.*

Practice the relaxation procedure daily until you have the control you desire.

## Cue-Controlled Relaxation

After you have become familiar with the relaxation procedure you should begin practicing cue-controlled relaxation. While relaxing each muscle group, say the word "relax" to yourself. By associating the cue word with a state of relaxation, the cue word, "relax," itself will eventually produce a relaxed state. Cue-controlled relaxation will not only be helpful for the control of math anxiety but also during other times of tension in your daily life. Each relaxation session should not last more than fifteen to twenty minutes.

As you learn to relax using these steps you will begin to realize that you have the power to relax your body and control your emotional state. As you practice the relaxation exercise you will recognize that relaxing is a motor skill which improves with practice. The more you practice, the better you will become. The more you practice, the more deeply and more quickly you will be able to relax. The converse is also true. If you don't practice you will not get better. Once you realize that the ability to relax resides within you there is little need to fear being overwhelmed with tension or anxiety because you will realize that you control your physical state.

If math anxiety has disabled you, remember that the disability you experience is your body getting itself ready to run or fight. Relaxation is the opposite, and mutually exclusive, state. During relaxation your mental capacity and potential are maximized. When you are relaxed you can think clearly; when you are anxious your thinking processes are inhibited.

# Chapter 4

# THE COGNITIVE ASPECTS
# OF MATH ANXIETY

In addition to the physical reaction of the body to anxiety, there is a cognitive, or mental, component at work as well. This mental and emotional aspect of math anxiety must be examined and modified in order to overcome the problem.

Anxiety is created by a person's expectations or thoughts about what is likely to happen. These thoughts can be expressed in words to oneself, mental pictures, or physical sensations. If a person believes, for example, that he or she is going to fail or appear stupid, that person will have an emotional reaction consistent with the expectation. The negative mental picture or self-talk created in one's mind produces the corresponding negative emotional reaction.

| | |
|---|---|
| **EMOTION IS THE BY-PRODUCT OF COGNITION** | **FEELING IS THE BY-PRODUCT OF THINKING** |

In reference to mathematics, if one believes oneself to be incapable in math, the expectation is that he or she will look stupid and feel humiliated when performing math. This causes the person to feel apprehension and anxiety about being in those situations which require performing mathematical computations.

Graphically, the process looks like this:

BELIEFS ⟶    EXPECTATIONS ⟶    ANXIETY ⟶    DISABILITY
*"I'm going*      *"I'll be humiliated"*     *Fight/Flight*    *Inability to think*
*to fail"*

## You Create Your Own Beliefs

People gather information about the world in which they live through their senses: their eyes, ears, nose, taste, and touch. Then they interpret the information they have gathered in light of previous experiences of similar nature. The process involves turning objective sensory data into subjective perceptual data. In other words, people attach meaning to what they see, hear, smell, taste, and feel. The meanings they attach reflect their own personal, subjective experiences. For example, it is not uncommon for ten people to witness an event and give ten different accounts of it because the event is interpreted from the perspective of each viewer's past subjective experiences.

After sensory data is given meaning, conclusions are drawn from the experience and form the basis of beliefs. These beliefs are also called opinions, attitudes, and self-concept. It is important to note that beliefs are subjective conclusions that have been drawn from subjectively interpreted experiences. And because beliefs are based on subjectivity, one individual's conclusions about an event might or might not correspond to others' conclusions about the same event.

Finally, people's **behavior** is a reflection of their **beliefs**, and their beliefs and behavior are always consistent. In other words, people act on the basis of what they believe to be true. When they act as if their beliefs are true they make their beliefs truth for themselves. A person's beliefs thereby become self-fulfilling. As many philosophers have stated in various ways throughout the centuries "As a man thinketh, so is he." Or to paraphrase in more modern terminology, "As a person thinks, so he or she is."

A diagram of the way people turn objective data into subjective beliefs looks something like this:

| SENSATIONS → | PERCEPTIONS → | CONCLUSIONS → | BEHAVIOR |
|---|---|---|---|
| Sight, taste, sound, odor | Meanings attached to sights, tastes, sounds, odors, etc. | Beliefs, attitudes, opinions | Action consistent with conclusions |

Now look at how the math anxiety pattern develops and is perpetuated:

| SENSATIONS → | PERCEPTIONS → | CONCLUSIONS → | BEHAVIOR |
|---|---|---|---|
| Seeing, hearing, feeling | Negative experiences involving math | "I'm bad in math; I'm going to fail math; I'll be humiliated if I try" (See other examples of math myths and self-talk.) | Apprehension, anxiety, fear, panic, worry, and avoidance when possible |

Take a few minutes right now to recall some of your previous math experiences and think about what conclusions you drew from them. Just what are your beliefs, opinions, and attitudes about mathematics and yourself in relation to math? Look at your behavior. Is it consistent with your beliefs? Are those beliefs and behavior self-enhancing or self-defeating? Are your beliefs helping you or hurting you? Are the beliefs true?

Avoidance is a typical behavior pattern of math-anxious people. By avoiding mathematics and math situations they reinforce the belief that they are incapable. In fact, each time they avoid math they are becoming more uninformed in that area. Then, every time math is approached they are less likely to succeed at it and their expectation of failure is again confirmed. If you have established a behavior pattern of math avoidance, you are probably **uninformed** in math but not **incapable**. Be aware of the difference between the two.

The pattern of math anxiety which many people **create for themselves** through negative experiences looks like the diagram on page 33. It is a self-defeating and self-perpetuating process. Math myths, negative self-talk, and visualizations, which will be subsequently discussed in this chapter, typically reinforce this math anxiety cycle.

Cognitive restructuring, which is explained in Chapter 5, is the solution for breaking the math anxiety cycle.

## Your Beliefs: Fact or Myth?

Some widespread beliefs about mathematics are particularly anxiety-provoking and self-defeating. They are

## THE MATH ANXIETY CYCLE

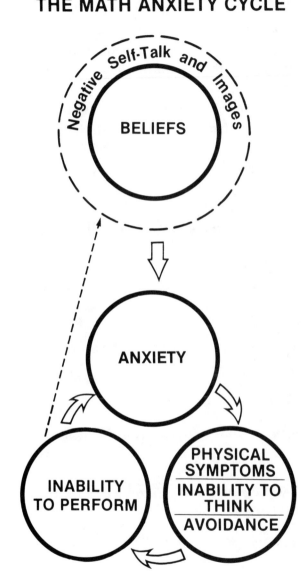

known as math myths because they are commonly held untruths shared by a great many people. These myths have little or no validity. Yet they are socially accepted as beliefs and are self-perpetuating because people act as if they are true despite the fact that they are not.

In their book, *Mind Over Math*, authors Stanley Kogelman and Joseph Warren identify twelve of the most commonly held math myths. These myths have caused people to fear and avoid mathematics. The twelve myths are listed below. Read them thoughtfully and note those which you believe.

1. *Men are better in math than women.*

2. *Math requires logic, not intuition.*

3. *You must always know how you got the answer.*

4. *Math is not creative.*

5. *There is a best way to do a math problem.*

6. *It's always important to get the answer exactly right.*

7. *It's bad to count on your fingers.*

8. *Mathematicians do problems quickly, in their heads.*

9. *Math requires a good memory.*

10. *Math is done by working intensely until the problem is solved.*

11. *Some people have a "math mind" and some do not.*

12. *There is a magic key to doing math.*

(For a more detailed discussion of each of these math myths, read *Mind Over Math* by Kogelman and Warren.)

Remember, these statements are not true; they are myths. Consider the degree to which you believe any of them. Then consider the consequences of believing them. If you find that you believe several of the myths you might gain some insight into your math anxiety. It is self-defeating to subscribe to any of them. Metaphorically, you might liken yourself to a runner trying to run a race with sand bags tied to his or her feet. The sandbags are an unnecessary handicap that inhibit the runner's chances of success in the race. Similarly, math myths can only hinder you if you believe them.

## Self-Talk:
## The Tape Playing in Your Head

The way we talk to ourselves is a reflection of our beliefs. We often interpret what we are experiencing by talking to ourselves silently, through self-talk. Self-talk is a monologue that we carry on with ourselves, often like a tape recording playing in our head. The same tapes play over and over.

---

**FEELINGS ARE THE DIRECT RESULT
OF WHAT WE SAY OR VISUALIZE TO OURSELVES**

---

Through self-talk we interpret our on-going experiences in light of our beliefs. For example, we say things like "you'd better hurry or you'll be late," "look at the absurd clothes on that person," "I hope the teacher doesn't call on me, I'd be so embarrassed," or "I will never be able to do math." We interpret to ourselves what is going on around us and what we ought to do about it. Our messages to ourselves are of course based on our beliefs and are often statements about our expectations. If we say threatening things we will feel anxious.

---

### ANXIETY IS PERPETUATED THROUGH NEGATIVE SELF-TALK AND THREATENING IMAGES

---

Most people have a few favorite lines that they say repeatedly to themselves or favorite visualizations that they often bring to mind. If these favorite lines or pictures are threatening, they will experience tension regularly. It is like repetitious tapes playing into a set of earphones or a video cassette playing a scene repeatedly on a television screen.

The next time you are experiencing math anxiety, **stop** whatever you are doing and listen to the monologue going on in your head. Chances are great it will sound like a tape that you have played over and over since childhood. Perhaps it is a videotape complete with images and scenes. You will discover that the emotions you are experiencing are consistent with the messages on your tape. These "recorded messages" are the roots of the anxiety you experience.

Listed here are a few of the most common math anxiety-producing tapes. Most of these messages were taken from *Mind Over Math* by Kogelman and Warren.

*"Everyone knows what to do, but me."*

*"I don't do math fast enough."*

*"I'm sure I learned it, but I can't remember what to do."*

*"I knew I couldn't do math."*

*"I don't have a math mind."*

*"I got the answer but I did it the wrong way."*

*"This may be a stupid question but..."*

*"It's too simple."*

*"Math is unrelated to my life."*

*"I should know that...it's obvious."*

*"That is a simple problem. If I can't do it I must be stupid."*

*"All I have to do to learn math is to work hard."*

*"It's no use—I'm going to fail."*

*"I'm going to make a fool of myself."*

*"People are going to laugh at me."*

*"The teacher is going to find out just how stupid I really am."*

If you stop to analyze the tapes you are playing in your head you will most likely find them to be completely absurd in that they have little or no external reality. Not only are they absurd but they are self-defeating because they create an emotional state consistent with the tape. Remember, the emotional/physical state creates the state of inability. Imagine trying to pass a math test with a live tape recording of your messages playing loudly in your ear!

## Visualizations:
## The Pictures in Your Mind

Some people use mental pictures, or visualizations, more than words to interpret what they experience. Like self-talk, our visualizations are an interpretation of current experiences in light of our beliefs. If we visualize something threatening, such as imagining ourselves flunking a math test, we will feel anxious. The negative image creates the corresponding negative emotion.

Become aware of your visualizations in relation to math. They may be daydreams and "videotapes" which perpetuate your math anxiety. Below are some examples of visualizations which cause anxiety. As you read them be conscious of your own mental images.

*Father or mother impatiently "helping" you with your math.*

*The harsh-speaking teacher whose intention was to embarrass students in front of the class.*

*Sitting in math class not understanding what is being taught.*

*Feeling stupid and wondering if everyone around you in class thinks that of you.*

*Visualizing yourself failing a math test.*

*Seeing yourself flunking out of school.*

*Visualizing yourself not being able to graduate.*

*Making a monumental error in a customer's account.*

Just like self-talk, your negative visualizations are self-defeating because of the emotional state that they create. It is important to restructure these "videotapes" so that you no longer allow them to create a state of inability.

## Childhood Recollections Influence the Present

Many people can pinpoint the onset of their math anxiety with a single negative math experience or a series of negative events from their past relating to math. In many cases, even experiences that happened in childhood decades ago still have a major impact on one's adult thoughts and actions years later. In fact, the impact of those events often seems irrational in light of one's present life and maturity, and yet it does indeed persist. The explanation lies in understanding how childhood memories are acquired.

Highly charged emotional experiences and conclusions regarding those experiences are recorded, as well as recalled, from the maturational level of the child who had the experience. The reality of the experience as well as the intensity and meaning of the event are remembered **as they were experienced as a child** rather than as they would be experienced if the event happened to the person as an adult. That is why a childhood math trauma from the past has so much effect on a person in the present, even though from one's current perspective the experience seems insignificant.

The way to change emotions attached to past experiences is to follow the restructuring procedure for visualizations.

## The Sad But True Story of Carl

Here is an example of how one person, Carl, developed his pattern of math anxiety and avoidance. He (1) had a negative experience from which he developed some beliefs; he then (2) acted on the basis of his conclusions; which (3) validated the original, although irrational, conclusions.

Carl was an eighth grade student who had been having some difficulties with mathematics. In order to foster competition and increase motivation, Carl's math teacher decided to seat the students in the classroom on the basis of their math test scores, with the best students sitting at the front of each row of desks and the poorest ones at the back. Early in the year, Carl contracted a serious case of the flu and missed ten days of school. The first day he returned to school his class had a math test. He did very poorly on the test and was moved to the end of the last row of desks. Carl was very embarrassed about his test score and, to make matters worse, some of his classmates teased him about it.

By the end of the week Carl had not caught up on the work he had missed during his absence and as a result did not fully understand the math problems the class was studying. Again Carl did very poorly on the exam and remained at the end of the last row.

By this time Carl was feeling inadequate, his self-esteem was shaken, and his confidence had decreased. He blamed his plight on his own inability. Nevertheless, Carl decided to give the next test his best effort. He inflated the significance of the test to the point of "all or nothing" importance, and studied as hard as he could. Because the test was so important to him, he felt a lot of anxiety by test time.

Unfortunately, Carl could not work the first problem on the test and found himself beginning to panic. He became very anxious and was unable to think clearly. It is easy, of course, to guess the rest of the story: Carl gave up and remained at the end of the long row of desks for the remainder of the year, the poorest math student in the class. Fortunately, he was doing quite well in all of his other classes that year; it was only mathematics that gave him difficulty.

Carl drew several conclusions from that year's experience: (1) that he was incapable in math, since he did try hard but simply could not do the work; (2) that he must not have a "math mind" since he was doing well in his other classes but was the worst student in his math class; (3) that it was obvious to everyone else that he was incapable in math; and (4) that when his math work was evaluated he would be publicly humiliated. Furthermore, he concluded that (5) taking any math course would result in failure and ridicule, and that (6) he had better avoid math when possible.

Carl's later actions were based on his conclusions. When he reached high school the next year he had the option of taking pre-college algebra or business arithmetic. Based on his beliefs about his inability in math, Carl chose business arithmetic and avoided higher math altogether. Note that it did not make any difference whether Carl's beliefs were rational or irrational because he acted in a manner that validated his beliefs: he did not take the pre-college course. By not taking algebra, Carl did not learn any pre-college math, which also confirmed his original belief that he was incapable and that he did not have a "math mind."

As Carl's case of math anxiety developed, an analysis of his self-talk would undoubtedly reveal self-defeating messages such as these:

*"I can't do math and I'll never be able to do it."*

*"Everyone will know that I'm stupid when it comes to math."*

*"I'll fail if I try to do math I always have."*

*"If I can't pass this test or this course I will never be able to pass any math course."*

*"Math is too hard for me."*

Carl, like many math-anxious students, fell into a cycle which reinforced his negative beliefs about his ability to perform in math. From the eighth grade forward, he avoided math when possible and became less and less informed about the subject.

## Math-Anxious Mary

The case of Mary is another example of how a previous negative experience can be transformed into current anxiety.

Mary is the wife of a prominent business executive; together they maintain a very active social life. One of her husband's favorite social activities is playing cards. Mary, however, feels very anxious when she is required to play. She avoids card games altogether whenever possible, much to her husband's dismay.

A glance at Mary's childhood reveals the roots of the anxiety she associates with card games. In the third grade, Mary's teacher would walk up and down the aisles of the classroom requiring students to stand as she gave them an arithmetic problem to solve aloud. If the students gave the right answer they could be seated and the teacher would move on to the next pupil, whom she selected at random. If, however, the students did not give the right answer, the teacher would strike their knuckles with a ruler before continuing on to the next student with the same "simple" arithmetic problem. Mary was terrified that she would get the wrong answer, be punished in front of the class, and experience embarrassment and humiliation. Because Mary was terrified, her ability to do "simple" arithmetic in her head was diminished and the thing she feared most usually happened: she got the answer wrong, was struck on the knuckles, and felt embarrassed and humiliated.

Consequently, as a third grader Mary drew several erroneous conclusions about herself in relation to math. Mary concluded that (1) she was very poor in math; (2) she ought to be able to do math "in her head" and (3) do it instantaneously; and (4) if she attempted to figure math in her head she would probably fail to perform adequately and (5) be publicly humiliated and suffer a loss in self-esteem.

It has been thirty years since that experience, but Mary still avoids math whenever possible, especially in public. Playing cards makes her feel as if she is going to make a fool of herself, embarrassing both her husband and herself because she cannot do "simple" arithmetic quickly and in her head.

Note that Mary had her unfortunate experience as a nine-year-old pupil. From that experience she drew conclusions about her ability and the consequences of her inability. Because she believes her conclusions to be true she acts in a manner that validates them and feels better when she avoids situations that involve math. The avoidance behavior is thus reinforced, which in turn increases the likelihood that the behavior will happen again in the next similar situation.

## Summary: The Math Anxiety Cycle

Math anxiety which is experienced in the present has its roots in the past. The victims have drawn conclusions or developed beliefs from those experiences (which are often forgotten) that indicate there is likelihood of emotional or physical threat. They visualize these threats or use threatening self-talk which portrays their expectations based on their beliefs. This elicits the vigilance or fight-or-flight reaction, which diminishes their ability to think clearly and in turn reinforces the original belief that they are inadequate in math. To compound the problem, because the physical state of anxiety is emotionally and physically unpleasant, they try to avoid it whenever possible. Avoidance causes them to be **uninformed** which they interpret as being **incapable**.

# Exercise 4-A

Describe any previous negative math situations you have experienced and how you felt at the time.

_____

_____

_____

_____

_____

_____

_____

_____

_____

_____

In the situation(s) described above, what conclusions did you draw about math or about your own ability in math?

_____

_____

_____

_____

_____

_____

_____

_____

_____

_____

## Exercise 4-B

Read the math myths listed on page 34. In the space below, copy those math myths that you have grown up believing.

_____

_____

_____

_____

_____

_____

_____

Describe your current ability in math.

_____

_____

_____

_____

Is the description above positive? negative? rational?

Have you ever avoided math courses or math situations?
_____ yes _____ no

If yes, describe the circumstances.

_____

_____

_____

_____

_____

_____

## Exercise 4–C

Analyze your own negative self-talk. At the moment when you are experiencing math anxiety, what are the negative messages (like a tape recording inside your mind) that you are telling yourself? Write them below. (See examples on pages 36–37.) Put an asterisk (*) next to those that you use repeatedly.

_____

_____

_____

_____

_____

_____

_____

_____

_____

Describe any negative visualizations (daydreams or "videotapes") which perpetuate your math anxiety. (See examples on page 38.)

_____

_____

_____

_____

_____

_____

_____

_____

_____

# Exercise 4-D

Describe the difference between being uninformed and being incapable.

_____

_____

_____

_____

_____

_____

_____

# Chapter 5

# COGNITIVE RESTRUCTURING

The remedy for dealing with the cognitive aspect of math anxiety is cognitive restructuring. Cognitive restructuring is a process of (1) raising to a conscious level the relationship between self-talk/images and feeling; (2) examining the consequences of irrational self-talk/images; and then (3) replacing irrational, self-defeating self-talk/images with rational self-talk/images; which results in (4) a change of feelings.

As was stated earlier, what one feels is a by-product of what one thinks. The philosopher, Euripides, said that the meaning of an event lies not within the event itself but rather within the person who experiences, then interprets, the event. Such is the case with a person who experiences math anxiety. The anxiety is not caused by the mathematics but rather by the meaning that the individual attaches to mathematics. The meaning that is attached to mathematics is symbolized in the individual's mind by words and pictures.

What one believes to be true is what one experiences as truth. If people believe that they are incapable in mathematics, chances are very good that they will perform poorly, thus demonstrating inadequacy, in mathematics. If you tell yourself or visualize that you are going to appear stupid or look foolish then you will have an emotional reaction which is consistent with that message. The emotional reaction will be one of anxiety because the messages are threatening. Cognitive restructuring is a process of changing those negative, self-defeating messages and images to rational ones in order to change the corresponding emotions from negative to positive.

## The Cognitive Restructuring Procedure

On a sheet of paper, make a chart with three columns on it. At the top of the first column, write FEELINGS; at the top of the second column, write SELF-TALK/ IMAGES; at the top of the third column, write RATIONALLY RESTRUCTURED MESSAGES. (You will use the first two columns in Steps 1 and 2. The third column will be used later, in Step 3.)

Your chart should look like this:

| FEELINGS | SELF-TALK/IMAGES | RATIONAL RESTRUCTURED MESSAGES |
|---|---|---|
|  |  |  |

# Step 1

Carry this chart around with you for a few days. When-ever you experience tension, worry, or anxiety, do the following:

*Write a description of your emotional and physical feelings in the* **FEELINGS** *column on the paper.*

*Then ask yourself what you are saying to yourself or visualizing that is threatening or causing the worry, tension, or anxiety which you recorded in the first column.*

*Write whatever you are saying to yourself or visualizing in the* **SELF-TALK/IMAGES** *column.*

As a result of recording your feelings and the corres-ponding self-talk and images on your chart, you will be developing a list of your own anxiety-producing messages.

Here is an example of what a self-talk/images log should look like:

| FEELINGS | SELF-TALK/IMAGES |
|---|---|
| 1. *My stomach is churning and I feel nauseous; I am tense and anxious.* | 1. *"I'm going to forget the information I studied for the test; therefore, I'm going to fail; therefore I'm not going to be able to achieve my goals and my parents are bound to reject me."*<br><br>*I see myself in math class failing the test; my parents are looking at me with great disappointment; I hang my head in shame.* |

| FEELINGS | SELF-TALK/IMAGES |
|---|---|
| 2. *My hands are cold and clammy and I'm perspiring; I feel nervous.* | 2. *I've just been handed the lunch check for a group of people and I'm expected to calculate each person's total. I see myself taking a long time figuring the amounts and then they all tell me the totals are wrong; they look shocked that I can't do it right. "I'm going to look like a stupid fool."* |

The typical thing that most people do is to have a negative thought about failing or appearing stupid, have the corresponding negative feeling, then immediately suppress the thought so that it never really becomes conscious and consequently cannot be rationally evaluated. The process repeats itself over and over. The person has the negative thought, the negative feeling, then suppresses it; has the negative thought, the negative feeling, then suppresses it.

The desired outcome of making the list of feelings and self-talk/images is to raise to a conscious level the anxiety-producing messages you say to yourself or visualize in order to break the supression cycle. By raising the messages to a conscious level you will then be able to see how they are making you feel anxious and defeated.

## Step 2

Once you have a list of self-talk statements and images, ask yourself questions about each one:

*"Is the message true?"*

*"Is the message helping or hurting me?"*

*"Is the message self-enhancing or self-defeating?"*

*"Is the message getting me what I want from life?"*

Answers to these questions will allow you to see the irrationality of your thinking as well as how it is causing you to feel anxious and defeated.

## Step 3

For the next step of the restructuring process, the self-talk statements and the images in the second column of the log should be separated because they are each dealt with differently. Follow Step 3 for the self-talk statements and Step 4 for visualizations. Use one or both methods for each situation on your list, according to whichever restructuring exercise works better for you.

*For each irrational self-talk statement in the second column, prepare a rational statement to replace it. Write each new message in the third column.*

*If you are saying to yourself that you are going to fail, write a message that says you are going to be successful and capable. Or you might write a statement that says how well you do on your math test will be determined by how well prepared you are, and that once you learn the information you will be able to demonstrate that you know the information. Again, for each self-talk message in the second column on your paper you should write a statement in the third column which is more rational and less self-defeating. In doing so you will turn a subconscious, self-defeating, emotional process into a conscious, helpful, rational process.*

You will discover that as you write and verbalize the rational statements, your feelings will change to be consistent with the new message.

Here is an example of rational restructured messages written to replace irrational self-talk:

| **SELF-TALK** | **RATIONAL RESTRUCTURED MESSAGE** |
| --- | --- |
| 1. *"I'm going to forget the information I studied for the test."* | 1. *"How much I recall will depend on the completeness of my study technique."* |
| *"I'm going to fail."* | *"I'm going to do the best I can do, and that's all I can do."* |
| *"I'm not going to be able to achieve my goals."* | *"It's not a matter of whether I achieve my goals but rather when I will achieve my goals; but I will achieve them."* |
| *"My parents are bound to reject me."* | *"My parents' love for me is not based on how successful I am; my parents' love is unconditional."* |

| SELF-TALK | RATIONAL RESTRUCTURED MESSAGE |
|---|---|
| 2. *"I'm going to look like a stupid fool when I can't figure out how much each person owes on the lunch check."* | 2. *"I'll get out a pencil and paper and figure it out; I can add and subtract; my human worth doesn't depend on being able to do arithmetic in my head."* |

## Step 4

*For each visual image noted in the second column of your log, visualize the scene as if it were a home movie filmed in black and white with a 16mm camera and projected on a screen in front of you. Watch your movie several times with the changes described below.*

*Imagine you are sitting in a movie theater in the back row. Play your move on the screen and watch it from start to finish. The beginning and ending must be points in time when everything is perfectly okay and you feel calm and serene. The disturbing, anxiety-producing situation should be in the middle of the movie. This may require you to create an ending for your movie so that when you watch it you play through the difficult part until you feel okay again.*

*Now make a melodrama out of your movie, much like the silent movies of the past. Have various people play villains, have a victim in distress, dress the characters appropriately, speed up the action, and have music playing in the background. Play it one to three times in this different form while you watch from the back row seat of the theater.* [1]

With this exercise your thoughts will become concrete images and you will dissociate yourself from their emotional impact. You will demonstrate control over your thoughts by changing their form and content, thereby altering the emotion associated with it. Remember, emotion is the by-product of cognition.

## Perpetuating Feelings of Inadequacy

Many of us cultivate feelings of inadequacy when, by most objective criteria, we are quite adequate. In reality, the human condition is one of both adequacy and inadequacy. At times we are brilliant and at times we are absolute fools; most of the time we function somewhere in between. Since both adequacy and inadequacy characterize all of us, we have to choose which one will be given our attention and energy.

Consider how you maintain your feelings of inadequacy when by most standards you are perfectly adequate. Although you do not have to look very far to find inadequacy in yourself, the converse is, fortunately, also true. In many things you are perfectly adequate. You must choose which one to focus on most.

Some people spend their time and energy worrying about appearing inadequate. They typically put a great deal

---

[1] This exercise is taken from *Using Your Brain — For a Change*, by Richard Bandler.

of effort into trying to protect themselves from others finding out that they are inadequate and maintaining an elaborate system of defenses that supposedly protects them from being exposed. The problem with this point of view is that it has self-protection as its central theme and that means that the person must constantly worry about being exposed. Constant vigilance about being exposed means constant anxiety.

A far healthier life perspective would be one of self-acceptance. Rather than trying to protect yourself from appearing inadequate, simply acknowledge that many times we are all less than adequate for the task at hand and just go on and do the best job you can do at the time. What you say to yourself only determines how you feel about what you are doing. It does not have the slightest thing to do with your ability. All you can do is to give a task your best effort. All you can do is all you can do. Why not just accept that as well as simply accepting yourself as you are.

First, it is important to acknowledge that it is impossible for you to be objective about yourself and therefore impossible for you to evaluate yourself objectively. Secondly, recognize that while you might **feel** inadequate it does not necessarily follow that you **are** inadequate. Thirdly, remember that to be human is to be inadequate in some or most ways. Fourth, accept yourself as you are. Fifth, focus your attention on the positive.

A good exercise for reversing the tendency to maintain feelings of inadequacy is to practice talking to yourself in the mirror, telling yourself that you are capable. You say this sounds crazy? Certainly not! Crazy is when you say self-defeating, irrational, and absurd things to yourself that cause you to become emotionally and cognitively disabled.

Try being kind, supportive, and encouraging to yourself. Treat yourself as you might treat a good friend. As you begin to do that you will discover a change in the way you feel. You will discover yourself to be a less anxious person.

In summary, the main purpose of cognitive restructuring is to help individuals see themselves as they are, both positive and negative, capable and incapable; to accept themselves as they are; and with this insight, to think and act **rationally**, which results in "feeling okay".

Cognitive restructuring places the responsibility for both the cause and resolution of math anxiety squarely upon the shoulders of the individual suffering the disorder. Through the exercise of keeping a self-talk log one can see how the individual creates and maintains anxiety. Furthermore, those who use this strategy will find themselves not the passive victims they once believed themselves to be. They — and you — can now take an active role in understanding and eliminating unwanted anxiety.

One final word on cognitive restructuring: knowing how and why it works is not enough; **it will not work unless you do it.**

## Exercise 5-A.
## The Cognitive Restructuring Procedure

1. Refer back to your answers in Exercise 4-C. In the first column, on the facing page write out separately each negative self-talk message or visualization that you use.

2. Then, for each irrational self-talk statement in the first column, write a rational statement in the second column to replace it. (See pages 50–52 for examples.)

3. For each visual image in the first column, visualize the scene as if it were a home movie filmed in black and white and projected on a screen in front of you. Watch your movie several times in different modes, e.g., as a melodrama, as an opera, backwards, in slow motion, fast forward, etc. (Refer to pages 52–53.)

# Exercise 5-A. (con't)

| Self-talk/Images | Rationally Restructured Messages |
| --- | --- |
| | |

# Chapter 6

# THE PSYCHOBEHAVIORAL ASPECTS OF MATH ANXIETY

The third component of math anxiety, the psycho-behavioral aspect, concerns cases in which a person's behavior seems to be inappropriate to the circumstances of a particular situation, that is, the behavior is not relevant or appropriate to the stimulus.

For example, one would not expect to elicit the vigilance and fight-or-flight behavior by the mere sight of a math classroom, teacher, or piece of paper with "Math Test" written across the top. Nor would one expect to be stricken with panic when handed a check for a group having lunch together and asked to figure out how much each person owes on the bill; or when handed a frequency distribution of previous sales within a company and asked to comment on it; or when asked to figure the simple interest on a loan. But over and over again people are stricken with panic when confronted with a situation involving the performance of mathematics, particularly when others are watching. These situations, by all external appearances, would seem to be non-threatening. Yet when confronted with particular math situations people are often victimized by panic and act inappropriately, or counter to what one might logically expect.

In order to understand why the inappropriate and often unwanted behavior occurs one must understand the principles of classical conditioning, stimulus generalization, transfer of conditioned emotional responses, and avoidance conditioning. While this chapter of the book is rather technical, these principles need to be explained in order to provide an understanding of the psychobehavioral components of math anxiety.

## Classical Conditioning

Classical conditioning is a process whereby two stimuli become paired together, and evoke a particular reflexive response. Conditioning results when a neutral stimulus is paired a number of times, or once very strongly, with another stimulus which is already directly wired to a reflexive behavior. Often the response does not seem related to the stimuli, such as when a person feels a sense of panic in a math situation which on the surface does not represent anything particularly threatening. In this case math has been linked with threat and with the biological reaction of fear.

Certain stimuli have biological and behavioral consequences which do not have to be learned but are evidently provided genetically. Reflexes are one such class of unlearned responses to specific stimulation of sense receptors. Salivation, eye blink, postural reflexes, and eye tearing, are examples of these "automatic" or reflexive responses.

Fight-or-flight is also an automatic response, caused by the stimulus, threat. Such responses are called unconditioned responses because they do not have to be learned; they just happen reflexively.

With reflexes, there is virtually a perfect correlation between the stimulus and the unconditioned response since the very survival of the organism might depend upon instantaneous reliable response. This response occurs "automatically" without thinking or learning. When there is no built-in mechanism for protection, each individual must learn what events and situations are potentially dangerous for him or her personally. The body then establishes a reflexive response for those situations. For example, a person who was bitten once in the past by a dog might always react to strange dogs with the same fear response.

On the subject of classical conditioning, introductory psychology textbooks usually cite the pioneering research of Ivan Pavlov. What Pavlov found was that when he presented meat powder to his dog, the dog reacted with the unlearned, automatic response of salivation. It was not long before other stimuli occurring at the same time — like sight of food, or merely the sight or sound of the experimenter —also became capable of eliciting salivation. Pavlov, as a result of these observations, decided to see if he could elicit an unconditioned reflexive behavior (salivation) by introducing a previously neutral stimulus (a bell).

Essentially Pavlov paired ringing a bell with food. He rang the bell, then gave the dog its food. The food caused the dog to salivate. He rang the bell and fed the dog repeatedly, and each time it heard the bell and tasted the food the dog salivated. Then Pavlov rang the bell but did not give the dog food — and the dog salivated. Amazingly, the neutral stimulus, a ringing bell, had caused a response, salivation, which was inappropriate to the stimulus. So the dog had "learned" an "automatic" or reflexive behavior which had no real relationship to the stimulus.

Much math anxiety is learned in the same manner. Stated more precisely, the biological anxiety reaction, the secretion of hormones and the biological process (the fight-or-flight reaction) that is set in motion by those hormones, is a conditioned response because there is nothing in the environment from which it would be appropriate to run from or to fight. Biologically speaking, math anxiety is a behavior which has been "learned" on a subconscious, automatic, reflexive level by pairing previous experiences which were painful with the activity of mathematics. The activity of mathematics then becomes the conditioned stimulus which elicits the conditioned response, anxiety.

What has occurred in such a case is a stimulus substitution: the functions of the original appropriate stimulus are transferred to a new, neutral stimulus. It should be noted that in this process neither stimulus is under the person's control.

The diagram on page 61 depicts the classical conditioning process.

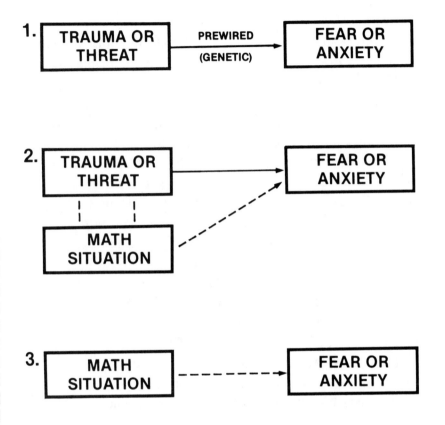

## Stimulus Generalization

During the early stages of the development of a conditioned response, many signals similar to the primary one will evoke the response. This phenomenon, called stimulus generalization, operates in a sense to open the organism to all stimuli which are related to the conditioned stimulus. It guarantees that a response will be made to a wide range of signals or stimuli, any of which might be the true stimulus. With additional experience, the organism responds only to stimuli which are increasingly similar to the true stimulus.

To continue with the example of the biting dog, the victim might at first fear all strange dogs, then learn to distinguish between growling dogs and friendlier ones which approach wagging their tails and are not aggressive in manner. As the victim encounters more non-biting dogs, the fear response is narrowed down more accurately to the true stimulus.

The tendency to "confuse" stimuli (stimulus generalization) is most pronounced when stimuli are very similar to each other, like situations involving math classrooms or math processes. One or more highly charged negative experiences with math can be generalized in such a manner to elicit anxiety or other negative emotions in all similar situations. For example, do you dislike all math situations, all math teachers, all math classrooms? If the answer is "yes" to any of these, chances are good that you have generalized your negative experiences (stimuli) to all similar stimuli (similar situations).

The story which follows (already familiar to anyone who has read an introductory psychology textbook) will help to further explain how stimulus generalization occurs.

## Transfer of Conditioned Emotional Responses
## or
## "The Unfortunate Tale of
## Little Albert and the White Rat"

Little Albert started out as a healthy, stable, rather unemotional baby. He never reacted fearfully to any of the test situations devised by the experimenter. His response to the succession of objects suddenly thrust upon him was to reach and play with them. The objects included a white rat, a rabbit, a dog, a fur coat, a ball of cotton, and some masks. However, Albert did startle and go into a fit at the unexpected sound of a steel bar striking a loud blow just behind him.

At the age of eleven months, when the white rat was presented and Albert reached for it, the steel bar was purposely struck behind him. After two such experiences, the baby was whimpering. A week later when the rat again appeared on the scene, Albert had learned his lesson — he withdrew his hand rather than touch his old playmate. At that point systematic respondent conditioning was started in order to establish a strong negative emotional response to the white rat. For seven trials, rat and startle noise were paired. Then when the rat was presented alone, Albert began to cry, turned, fell over, and crawled away with all his might. About a week later, this same fear reaction had generalized from the white rat to the friendly rabbit. The dog frightened him, the fur coat made him cry, he pulled away from the cotton ball, but saddest of all, he was pronouncedly negative when shown a Santa Claus mask. No such fear was shown to blocks or objects which did not share the apparently controlling stimulus dimension of "furriness."[1]

By now, the reader might have begun to visualize any number of situations that involve the pairing of fear and mathematics. The fear might be fear of failing, fear of looking stupid in front of friends, fear of rejection, or fear of punishment, to name just a few. It is common for fear associated with a specific math situation or set of circumstances to be transferred to similar situations and circumstances. On the surface the emotional response to the new situation or set of circumstances is inappropriate. The anxiety the person experiences was appropriate to the original situation but is now transferred to a new situation in which the reaction is no longer rational or appropriate.

## Avoidance Conditioning

Behavior which is reinforced is much more likely to recur than behavior which is not reinforced. Positive reinforcement is an event which happens after an action and increases the likelihood that the behavior will be repeated. An example of positive reinforcement is rewarding a child by giving him or her a piece of candy, or praise, or affection, for closing the door after entering a room. The next time the child enters the room chances are increased that the child will again close the door.

Negative reinforcement likewise increases the likelihood that behavior will be repeated. However, negative reinforcement is an escape from something negative. For example, if a child is punished for misbehavior by being

[1] Watson, J.B. & Rayner, R. "Conditioned Emotional Reactions," *Journal of Experimental Psychology*, 1920, 3, 1-14.

locked in a room alone and not being allowed to exit, avoiding or escaping the confinement would be a negative reinforcer; in this case the child performs the desired behavior in order to avoid the negative consequence.

Reduction of anxiety is a negative reinforcement. If a person can prevent a unpleasant situation by avoiding it or escaping from it the resulting reduction of anxiety is a negative reinforcement. After that occurs several times and the behavior becomes automatic it is called avoidance conditioning.

If a person experiences anxiety associated with being in a classroom where math is taught, then experiences a reduction of the anxiety by leaving the classroom, the person has experienced a negative reinforcement for this behavior and the probability of the avoidance behavior recurring is greatly increased. The behavior is the action of leaving the classroom; the negative reinforcement is the reduction of anxiety that results and the consequent feeling better that follows. After several similar situations in which a person reduces anxiety (and as a result feels better) by escaping or avoiding, the phenomenon of avoidance conditioning begins to occur. After the behavior is conditioned it becomes an "automatic" or "subconscious" habit. The person experiences a math-related situation, feels anxious, escapes or avoids the situation, and feels relieved. With each successive situation handled in this manner the behavior becomes more automatic and subconsciously entrenched.

## Nancy the Ninth Grader

Armed with the concepts of classical conditioning, stimulus generalization, transfer of conditioned emotional responses, and avoidance conditioning, consider the experience of Nancy as a ninth grade girl in an algebra class.

Nancy, a young and self-conscious teen, was called on by the teacher to go to the chalkboard and work an algebra problem for the class. On her way to the front of the room she heard several boys snickering and sensed that they were talking about her. She felt embarrassed and self-conscious but continued to the chalkboard. The teacher read an algebra problem to her which seemed vaguely familiar, but she could not quite remember how to get started on it. As Nancy stood at the chalkboard she felt as if all the eyes in the classroom were boring into her and the embarrassment she was feeling earlier was greatly increased. She began to feel hot flashes going through her neck and body and she started to perspire. Butterflies danced in her stomach and she felt short of breath. A "lump" appeared in her throat, making it very difficult for her to talk. Her heart began to pound and she felt like running, but she knew that would be inappropriate. She felt trapped with nowhere to run and began to panic. Nancy's thinking became a blur and there was absolutely no way she could begin to work the algebra problem. The teacher experienced great frustration because the problem was "very easy" and Nancy had worked several just like it on her homework assignment. As Nancy took her seat she felt humiliated and stupid, and was convinced that she had made a fool of herself in front of her classmates. To make things even worse the teacher had a boy go to the chalkboard after her; he proceeded to put the problem on the board and solve it with the greatest of ease. Nancy was sure that she would be the topic of discussion in the hallways between classes.

Nancy remembered this event for several years and avoided taking math for the remainder of her high school days. The mere thought of the humiliating experience gave her gooseflesh. Not only did Nancy avoid taking math classes, she avoided math in any form when possible. When a situation arose in which she had to do even simple arithmetic she would become anxious and unsure of herself. She always felt better when she had someone else do the arithmetic for her.

In Nancy's case classical conditioning occurred as math and threat were paired. Avoidance conditioning caused her to avoid math and feel better about it.

## Nancy as a Displaced Homemaker

As fate would have it, Nancy was forced by personal and economic circumstances to return to college at age thirty. She had long since forgotten the specific event in the ninth grade which ultimately led to her fear of mathematics. Nevertheless Nancy realized that most college majors and most professional jobs require knowledge of mathematics. She felt both incapable and uninformed; most of all she felt terrified. However, Nancy was aware of the necessity of knowing mathematics and for two semesters she enrolled in a beginning algebra class. She subsequently withdrew from the course each semester because she was so uncomfortable in the class that she "just couldn't think straight."

An armchair analysis of Nancy's situation reveals that she had a traumatic experience as a ninth grade student which she paired with mathematics. The trauma had nothing to do with the subject matter, mathematics. The incident primarily involved fear of appearing stupid in front of classmates, fear of experiencing humiliation, and fear of ridicule and rejection. The same thing could have happened while she was diagramming sentences for an English class. However, Nancy was in a math class and she paired embarrassment, ridicule, humiliation, and panic with math. Since the reaction, panic, was very strong it only took one time for the classical conditioning process to begin. The reaction, fear (increased blood pressure, perspiration, butterflies, lightheadedness, and inability to think), became a conditioned response associated with the conditioned stimulus, mathematics. The conditioned response was generalized to include all math-related activities and the anxiety was likewise transferred to all math-related activities. Avoiding math became a self-perpetuating behavior because it allowed Nancy to reduce the amount of anxiety she experienced when she thought about or was confronted with mathematics.

The final result of this process is that Nancy still irrationally fears math and continues to avoid it in college, even knowing that she needs it for her career preparation. Looking at Nancy's behavior from her college teacher's point of view, the fear and avoidance would appear inappropriate to the situation, illogical, irrational, and self-defeating. It is clear, however, that Nancy is a victim of math anxiety. If this pattern of avoidance is allowed to continue, she, like many other college students who have math anxiety, will remain uneducated or uninformed in this subject matter.

> **MATH ANXIETY IS
> A CLASSICALLY CONDITIONED RESPONSE
> THAT HAS BEEN GENERALIZED
> AND IS REINFORCED AND
> THEREBY PERPETUATED BY AVOIDANCE**

# Chapter 7

# DESENSITIZATION

Desensitization is prescribed for reversing the psycho-behavioral aspects of math anxiety. It consists of using the relaxation procedure in combination with mental images to desensitize a person to scenes which have been anxiety-producing in the past.

Systematic desensitization for anxiety is a behavioral therapeutic intervention strategy pioneered by Joseph Wolpe in the late 1950's and early 1960's. It deals with anxiety on a physical level. Systematic desensitization is a process that modifies the reflexive, automatic, or conditioned reactions by having the person learn new reflexive responses to previously anxiety-producing situations. The change is accomplished by pairing the physical state of relaxation with mental images of previously anxiety-producing situations (in this case math-related). With this pairing new responses will be learned on the reflexive or "automatic" level. The new responses will be relaxation and a sense of well-being associated with math and math-related activities.

In order for you to become desensitized you will need to develop a math anxiety "hierarchy" to use in the desensitization procedure. This consists of a list of math anxiety-producing scenes arranged in order of the degree of tension associated with each one. The first scenes are those which cause you some but little difficulty; the later scenes are those which cause you great difficulty.

## Developing Your Hierarchy

*Make a list of situations involving mathematics which cause you tension, worry, or anxiety. Rank each scene on your list according to the following five categories: Weak Anxiety, Mild Anxiety, Moderate Anxiety, Moderately High Anxiety, or Intense Anxiety. You should have at least two scenes in each category. If not, create some new scenes or modify an existing scene from an adjacent category so that it can be placed into the deficient category.*

*After you have at least two scenes in each category, assign them specific numerical values so that you can arrange them in a hierarchy ranging from very weak anxiety to intense anxiety.*

The following *subjective anxiety scale*, developed by Joseph Wolpe in addressing the problem of spacing of hierarchy scenes, may be helpful to you:

"Think of the worst anxiety you have ever experienced, or you can imagine experiencing, and assign to this the number 100. Now think of the state of being absolutely calm and call this 0. Now you have a scale.

On this scale how do you rate yourself at this moment? The unit is called a Subjective Unit of Disturbance *(SUD)."*

The five categories that you used to begin ranking the scenes in your hierarchy would fall under the following *SUDs* values:

|  | SUDs Value |
|---|---|
| Weak Anxiety | (1-19) |
| Mild Anxiety | (20-39) |
| Moderate Anxiety | (40-59) |
| Moderately High Anxiety | (60-79) |
| Intense Anxiety | (80-100) |

*Now rate the scenes of your hierarchy according to the amount of anxiety you would have upon exposure to them. Try to have 10 to 20 scenes in your hierarchy with no more than 10 points between each scene.*

It is important to have a smooth progression from a lower level of anxiety to a higher level of anxiety. If the differences between the scenes in your hierarchy are similar, and, generally speaking, not more than 5 to 10 *SUDs*, the spacing can be regarded as satisfactory. On the other hand, if there are, for example, 10 *SUDs* for scene eight and 40 *SUDs* for scene nine, you need to create some intervening scenes.

On page 74 is an example of a hierarchy of anxiety-producing situations related to taking a math test. The *SUDs* value assigned to each scene (as interpreted by the person who compiled the hierarchy) is given in parentheses.

1. *Sitting in a math classroom (10 SUDs)*

2. *Seeing the teacher come into the classroom (15)*

3. *Having the teacher assign a test date for an upcoming test (20)*

4. *Preparing for a test the night before the test (25)*

5. *Waking up on the morning of the test to remember that I have a test (30)*

6. *Driving to school to take the test (35)*

7. *Walking to the classroom to take the test (40)*

8. *Standing outside the classroom talking with friends before going in to take the test (50)*

9. *Being in the classroom just before the test (60)*

10. *Seeing the instructor come into the classroom with the test (65)*

11. *Having the instructor hand out the test (75)*

12. *Beginning to work on the test (80)*

13. *Seeing the clock on the wall during a test (90)*

14. *Finding a problem I cannot work on the test (95)*

15. *Turning in the test and walking out of the classroom (95)*

Note that although the scenes in the hierarchy above may appear to represent a sequence of events, this is not a requirement in developing a hierarchy. The order of the scenes is determined by the increasing degree of anxiety associated with each situation.

Although systematic desensitization was originally developed by Wolpe to be administered by a therapist, recent research has indicated that self-administered systematic desensitization is equally effective. Kahn and Baker, for example, compared conventional desensitization to "do-it-yourself" desensitization with college students with "subclinical phobias" and found it to be equally effective for that group.[1]

## The Desensitization Procedure

Now that you have developed your hierarchy of math anxiety-producing situations, use the following procedure to desensitize yourself to those scenes:

*Using the relaxation exercise in Chapter 3, relax as completely as possible.*

*Visualize a calm and peaceful scene.*

*Visualize scene number one from your math anxiety hierarchy, i.e., a scene in which you experience a slight amount of tension. Concentrate on relaxing as you visualize this scene.*

*Visualize the next scene from your hierarchy as you continue to relax. Spend about 10 seconds visualizing each scene before you proceed to the next. If you have difficulty relaxing in a scene, shift backwards to the previous scene in which you were relaxed. Visualize that scene until you are completely relaxed again, then return to the scene which caused you difficulty before. The second time around you*

---

[1] M. Kahn and B. Baker, "Desensitization with minimal therapist contact," *Journal of Abnormal Psychology*, 1968, 73, p.198. Similar results have been obtained by Repucci (1969); Phillips, Johnson, and Geyer (1972); and Dawsley, Rosen, Glasgow, and Barrera (1973).

*should be able to visualize the scene while relaxing. If you cannot relax with a scene, repeat the process of shifting back and forth between scenes until you can completely relax with each scene.*

*Continue until you can visualize each scene in your math anxiety hierarchy while at the same time maintaining a state of complete relaxation.*

The duration of each scene should be 5 to 10 seconds. The interval between scenes should be 10 to 20 seconds.

Spend no more than 30 minutes in this process once or twice a day until you become desensitized.

As you visualize scenes which previously produced anxiety you will find yourself becoming increasingly unresponsive physically and emotionally to those scenes. In other words, you will become "desensitized." As you become desensitized you will discover a change in the way you respond to those real life situations. With each relaxation and desensitization exercise you should feel less and less math anxiety.

If you find that your math anxiety has not diminished after several sessions, check to see if any of the following might be problems for you:

1.  Not being able to relax completely;

2.  Having too many *SUDs* between scenes in your hierarchy; or

3.  Not being able to visualize the scenes in your hierarchy clearly.[2]

---

[2]The author would like to offer one note of caution. Exposure, and prolonged exposure in particular, to a very disturbing scene can increase rather than decrease the amount of anxiety associated with that scene. In order for desensitization to occur, the person must be relaxed while viewing the scene.

There is one requisite for this type of intervention. In order for relaxation and desensitization to occur **you must do it!** Being aware of the intervention strategy or understanding how the strategy works is not enough. An analogy might be made to a physician's prescription. It is not enough to receive a prescription, or to have the prescription filled, or even to understand why it would be good for you to take the prescription; ultimately, to be helped you must take the medicine. Such is the case with relaxation and desensitization. It works only when you do it.

## Exercise 7-A.
## Developing a Hierarchy—Part 1

On the facing page, make a list of situations involving math which cause you tension, worry, or anxiety. Then rank each scene on your list according to the level of anxiety it evokes. Use the following categories:

Weak Anxiety

Mild Anxiety

Moderate Anxiety

Moderately High Anxiety

Intense Anxiety

You should have at least two scenes in each category. If not, create some new scenes or modify an existing scene from an adjacent category so that it can be placed into the deficient category.

# Exercise 7-A. (con't)

| Math Anxiety Scenes | Level of Anxiety |
| --- | --- |
| | |

# Exercise 7-B.
## Developing a Hierarchy—Part 2

Read about the subjective anxiety scale and the Subjective Units of Disturbance (SUDs) on pages 72–75.

Next, go back to Exercise 7-A and rank each of your anxiety scenes with a SUDs unit. Then, turn the page and copy your list as a final hierarchy.

In the space below, copy the list of scenes from Exercise 7-A in the form of a hierarchy by putting them in order from the lowest SUDs value to the highest.

# Exercise 7-B (con't)

| Math Anxiety Scenes | SUDs Value |
| --- | --- |
| | |

Now you have a hierarchy list to use for the desensitization procedure described on pages 75–77.

# Chapter 8

# OTHER PERTINENT ISSUES

## The Father/Husband Syndrome

There is a phenomenon prevalent among women who suffer math anxiety which the author has chosen to call the Father/Husband Syndrome. Although male children can also have the experience, the syndrome is more common among females.

Women who experience this syndrome often report that their father was a mathematician, engineer, accountant, scientist, or employed in some area of applied mathematics. They recount that their father helped them with their math homework as a child and as a result of this tutoring interaction they concluded that they were not capable in mathematics.

Here is a typical situation related by these women. The child asks her father for help with a problem from her math homework. The particular problem is difficult, indeed impossible, for the child to solve alone, which prompts her to seek help. The father is delighted because his daughter has asked for help in an area in which he feels he can be helpful. The father gives the child an instantaneous answer to the problem, which to him is very simple because of his math background. He then explains the "simple" problem to the child in words which are unfamiliar to her because she is functioning with the vocabulary and conceptual framework of an elementary school child. She does not grasp the "simple" explanation given by the father, who proceeds to give one explanation after another for the problem. After all, the father recognizes that the child has requested help from him in a subject area about which he feels very knowledgeable (a situation which may be a rare occurrence for many fathers.)

After several unsuccessful attempts to explain the problem, both the father and the child become frustrated. The father feels frustrated because the child has not comprehended what he considers a series of excellent explanations of the problem. At the same time, the child is frustrated that she has not understood what are supposed to be "simple" math solutions.

Multiply this incident several times and you see the roots of the Father/Husband Syndrome. The child concludes that she is inadequate in math because, after all, if she were adequate she would be able to understand her father's explanations. Unfortunately, the father sometimes draws the same conclusion.

Several spinoffs occur as a result of the father's frustration and the child's conclusions. (1) The child senses that mathematics is an area in which the father is all-knowing and in which (2) he will consequently be able to detect the child's weaknesses. (3) The child believes that if she persists in trying to do mathematics she will expose her inadequacies to the father and (4) that he will continue to be frustrated with her and (5) consequently possibly reject her. This vulnerability is too great for the child to risk.

Since the child wants to be pleasing and capable in her father's eyes, she avoids mathematics whenever possible because she wants to avoid the possiblity of rejection. As a result of the avoidance, she becomes mathematically uninformed but defines herself as incapable. Recognizing her own supposed incapability, she channels her talents and energies into other endeavors, such as language arts or music.

Both father and child often fail to recognize the disparity between their levels of vocabulary, experience, and cognitive frameworks. If the father is an engineer, for example, he has taken many courses in mathematics, including arithmetic, algebra I and II, college algebra, linear algebra, trigonometry, analytical geometry, calculus I, II, and III, and differential equations — all as prerequisities to being admitted to engineering school. Obviously there is a large difference between the child's and father's quantitative abilities and vocabulary.

Later, a woman who has established this pattern with her father will often choose a similar man as a mate. This is because people often choose spouses who have the characteristics they admire most in their parents or spouses who have characteristics which complement their own. For these reasons women who have defined themselves as incapable in math often marry men who are competent in that area. They commonly defer to their husbands in situations that require mathematical competence, thus perpetrating the childhood belief and the ritual associated with it. One consequence of this pattern is that husbands are trained to treat their wives as if they are incapable in math. This pattern also perpetuates the myth about men being better in math than women.

The Father/Husband Syndrome is evident among many women who have been away from school for several years and for one of many possible reasons return to college to develop some skills that will allow them to get a job or build a career. Because of the highly technical nature of the evolving world of work and the role of mathematics in that technical world, it is undesirable if not impossible to avoid mathematics when preparing for a career.

For women with the Father/Husband Syndrome who are returning to college, the self-defeating conclusion that they are incapable in math must be re-evaluated. They must learn that they are uninformed rather than incapable.

## Brain Damage and Math Anxiety

College students suffering from math anxiety typically want to believe that there is some neurological damage that prohibits them from doing math. Sometimes they have a conscious hope that there is something wrong with their brain that would explain "it," their math disability, once and for all. Such a diagnosis would allow them to give up after all these years with justification for doing so.

Based on a review of the psychological literature and on personal clinical experience, the author feels safe in stating that brain damage can be ruled out as the cause of math problems unless there is also evidence of brain damage in various other, non-math, problem-solving situations as well. If there is enough brain damage to interfere with learning and performing math, there is great likelihood that it also interferes with certain other tasks as well, such as sequential reasoning, writing, or learning a language.

# THERE ARE NO "MATH-ONLY" BRAIN CELLS!

Solving a math problem requires the same cognitive processes that most other processes require: recognition, recall, analysis, synthesis, and generalization. These the same cognitive processes necessary for reading a map, following a recipe, learning a foreign language, writing a theme, understanding comparative literature, and fixing an automobile. To believe otherwise is to believe in a self-defeating myth.

Students who "wish" that brain damage were the explanation for their math disability rarely actually seek a neurological checkup or psychological testing — usually because they know that there is no brain damage. If they were to have a neurological exam and discover that there is no brain damage, they would have to accept something far worse than brain damage: the fact that they have not lived up to their potential. One of their greatest fears is that they will do the very best they can do, really put themselves on the line, and still fail.

The purpose behind believing the "brain damage" myth is that it allows the individual to justify a lack of success. If people can justify their lack of success, then they don't have to try. If they don't try, they can't fail. If they don't fail, they avoid the humiliation of failure.

There are problems inherent in believing the myth, however. To do so insures that the person will never learn how to do math and will suffer all of the life and career consequences of mathematical illiteracy.

# Learning Mathematics is Hard Work

Many people label mathematics as "hard" when it would be more correct to say that learning mathematics is "hard work." Mathematical concepts that have been developed over the centuries are concepts that took brilliant people many years to formulate. These concepts are now taught in our elementary and junior high schools. Calculus, for example, is now offered in most secondary schools across the country. Nevertheless, these concepts are still difficult ones to grasp without some real effort.

Learning mathematics requires hard work and a large investment of time and energy. Unfortunately, many students have never been challenged with hard work or difficult concepts until they encounter mathematics. When they run into difficulty they compare math with other school subjects which have not been as difficult for them. They often erroneously conclude that they are not as capable in math as they are in other subjects, simply because other subjects are less time consuming.

It should be reassuring to learn that a majority of students find mathematics more difficult or time consuming than other subjects. It typically takes more time and concentration to understand the concepts of mathematics. In addition to understanding the concepts, students must learn to apply them. Each concept then provides the foundation on which a new concept is built.

People who have had difficulty with mathematics as children often avoid studying, learning, or, wherever possible, using math. Several things happen as a result of this avoidance. First, they have substantial gaps in their learning. To use an analogy, learning mathematics is much like building a block fence. It is started by pouring a foundation, then laying numerous rows of blocks, each row on top of the last. Children who are math avoiders most often have a number of key blocks missing.

The result of missing blocks is that it makes later concepts either very difficult or impossible to learn. When this happens, people often erroneously conclude that mathematics is "hard" and that they are incapable in math. In actuality they are uninformed in math rather than incapable. Of course their belief that they are incapable can be readily substantiated by the fact that the new math concepts that they are trying to acquire continue to be difficult or impossible to learn. And so the incorrect belief is perpetuated.

Secondly, people who perceive math as "hard" often develop some anxiety around the subject because they are afraid that they will be unable to perform adequately. After all, the harder something is, the greater the likelihood of not being successful at it. The anxiety prompts avoidance and avoidance causes "missing blocks" and feelings of inadequacy.

Thirdly, it is a general human tendency to choose to do the things we do well because there is much greater likelihood that we will get a pat on the back for being successful. We do the things we are likely to be rewarded for and avoid the things in which we are likely to be unsuccessful. Children who have learned, no matter how irrational the conclusion, that they are not good at mathematics, usually choose to spend their time in other activities, ones in which they believe themselves able to do well. These individuals invest their time and energy in developing skills in other areas, sometimes at the expense of developing skills in math.

To reiterate, the principal theme here is that learning mathematics is a challenging process, one that requires time, energy, commitment, and plain old hard work. In order for students to be successful in math they have to "roll up their sleeves" and work at it — even if all other subjects seem relatively easy for them. Students who are willing to work and persevere at math are almost always capable of being successful.

## "Math Is Not Relevant To My Life"

The above statement is voiced frequently by high school and college students. The author's response to this remark is "You can't know what you don't know." In other words, you cannot understand a subject or its relevance if you have not been exposed to it.

An example of the same attitude in the author's past is his memory of sitting in class one day in fifth grade and marveling at the fact that he could not think of anything he did not know. He even wondered why he should continue to go to school since, after all, there was nothing he could think of that he did not know. He was ignorant of the existence of further knowledge and of his own limitations.

"You can't know what you don't know" means that we cannot, for example, see life from a historical perspective if we have not studied history; we cannot view life from an anthropological perspective if we have not studied anthropology; and we cannot understand the relevance of mathematics to our lives if we are mathematically illiterate.

Understanding mathematics allows us to think in quantitative terms. It is very relevant to today's technical world. To believe that mathematics is irrelevant to life is to say that you are ignorant of its relevance because "you can't know what you don't know."

## The Drive for Perfection

The personality dynamics of the typical math-anxious person often includes perfection as a goal. As children many people grow up believing that if they could just be "perfect" or "act right" they would be acceptable to their parent(s). Children as well as parents often do not distinguish between "being" and "acting" right. Children define "right" as what they believe their parents want them to be and do; this becomes synonymous with the child's definition of perfection. Being perfect means acting in a way that is consistent with the wishes of the parent and significant others.

As children mature, their definition of perfection is internalized, which means that it becomes a part of their own adult belief system. Adults who suffer math anxiety usually believe that in order to be okay, lovable, or acceptable they must be perfect. This desire to be perfect is usually subconscious but can nevertheless be seen in the expectations and behavior of the anxious person.

Because there is no such thing as perfection among humans, perfection is impossible to attain. The point to be made here is that if perfection is the criterion by which people measure themselves they will always fall short and thereby experience **feelings of inadequacy**, which in turn creates anxiety.

The higher people set their goals the greater the likelihood that they will not achieve those goals, thus experiencing feelings of inadequacy and corresponding feelings of anxiety. When persons try — and fail — repeatedly to meet their standards of perfection, the logical result is that they will feel like failures. It then follows that people who feel like failures will also feel inadequate and will fear that others might ultimately discover their inadequacy, particularly in the areas of greatest vulnerability, such as mathematics.

The process can be visualized as follows:

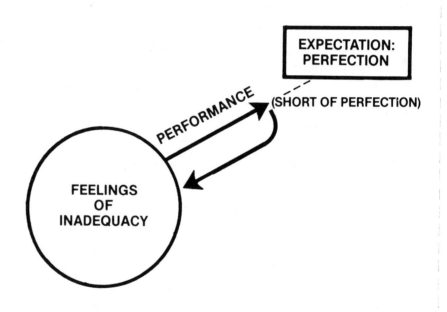

An example of this behavior is the perfectionist who makes a list of ten items to be completed on a particular day. At the end of the day, eight of the ten items are done. Rather than feel pleased about having accomplished 80% of his or her goal for the day, the perfectionist chooses to feel bad (inadequate) that two of the items on the list were not completed.

Feelings of inadequacy, and the resulting sense of failure, are based on unrealistic expectations and irrational conclusions regarding personal potential. The cure for this pattern is self-acceptance. It requires setting and living by rational goals and realistic expectations.

# Chapter 9

# EFFECTIVE STUDY TECHNIQUES FOR MATHEMATICS

This chapter is not intended to be a complete compendium of effective study methods. Rather, its purpose is to identify some key ingredients essential to the effective study of mathematics. A surprisingly large number of students do not know whether or not they have studied effectively or even if they have learned the material they have studied. After reading this chapter you should be able to tell whether or not you have learned the material you wanted to learn.

Most students know that it is important to work a number of sample homework problems. Sometimes working homework problems is the only thing that students do to prepare for a math test. They make the assumption that if they work problems over and over, learning will occur. They are only partially correct. This learning technique, repetition, is only a first step in the learning process. Repetition is important but repetition alone does not allow for anticipating slight differences between problems. Nor does it allow for making generalizations or inferences, nor for analysis of the problem, nor for synthesis of meaning. So students who have used repetition as the sole method of preparation for a math test often find themselves stumped on the test.

The problems from one assignment are usually similar to — but slightly different from — those problems on the assignments just before and just after. A math test will generally have several types of similar problems on it. It is because of the close similarities and differences between each group of problems that students need to use a multi-faceted study approach.

In order to perform math satisfactorily, one must **recognize** the nature of the problem to be solved and **recall** how to solve it. The study technique of repetition does provide the basis for recognition and recall, but the student must also be able to **contrast, analyze, synthesize**, and **generalize**.

## CONTRAST: Identify How Each Group of Problems Is Different Than Any Other Group of Problems You Have Done

It is essential that you contrast the similarities and differences among the problems to be solved. This means you should be able to verbalize, in your own words, how each problem is different from and similar to other problems in the units you have been studying.

## ANALYSIS AND SYNTHESIS: Explain In Your Own Words How To Do a Problem

Students often spend a great deal of time worrying that they are going to forget the material they have studied for a test. Rather than worrying (being anxious) about what you don't know, ask yourself how you can be certain that you do know the material for a test. The answer to that question is relatively simple: explain in your own words how you would solve each kind of math problem.

In order for you to explain verbally a concept, you must first have the information put together in your own mind before you can repeat it. That cognitive process is called synthesis. Have you ever felt that you knew something vaguely, yet as you explained it to another person the information became quite clear to you? If so, you have experienced synthesis. If you can explain how to do a math problem to another person, you know the material and you probably will not forget the information for a test. On the other hand, if you cannot explain how to work a problem, it is unlikely that you will consistently be able to get the problem correct on a test.

One of the keys to success in mathematics is understanding the terminology. Math has its own vocabulary and it is important for a student of math to be able to understand and to speak the language of mathematics. Be sure that you comprehend and can manipulate math terminology correctly.

## GENERALIZATION: Create and Solve Some Sample Problems of Your Own

If you know how a problem is similar to and different from other problems you have studied and you are capable of describing how to solve a problem, then you should be able to create new problems of similar nature. That cognitive process is called generalization and is the key to "owning" the information. If you can make applications of a problem-solving technique then you "own" that technique: it is yours and you will not likely forget it soon. When you are studying for a test, create a few problems of your own and solve them. Make sure that you create a number of problems for each variety of problems you are likely to encounter on the test.

Make the problems you create relevant to you and your life situation. Have you ever gone to a party and been introduced to someone of little or no importance to you, only to discover that ten seconds later you have forgotten the person's name? That is about how long any of us remember irrelevant information. On the other hand, suppose you were introduced to a person at the party who offered you a very lucrative job. Chances are very good that you would remember that person's name because that information is relevant to you. The principle here is that you will remember information that is relevant and you will not remember irrelevant information. If you make math problems relevant it will be easier to remember them.

## Putting It All Together: Preparing for a Math Test

1. *The first step in preparing for a math test is to read and study the textbook explanations of problems.*
2. *Work the sample problems in the textbook or on the worksheets the instructor has given you.*
3. *Explain aloud how to solve each type of problem.*
4. *Contrast aloud and in your own words the similarities and differences in the problems to be learned for the test.*
5. *Create a few sample problems of each variety you are studying and then solve them.*
6. *Explain how each type of process can be used to solve a problem you might encounter.*
7. *Create a sample test for yourself. Many textbooks have the answers to the odd numbered problems in the appendix. If your text is such a book, select a few sample problems of each type you are studying, scramble them and create for yourself a sample test.*
8. *Determine how much time you will have to complete the actual test. Select the number of problems you anticipate being on the test for your sample test. Set a kitchen timer, stopwatch, or alarm and test yourself. You will determine in this exercise how long it will take you to work each type of problem and how many problems you are likely to complete in a given period of time.*
9. *Review the problems on which you make errors.*

If you have demonstrated that you can work the problems on the test, you are adequately prepared to take a test. If you are prepared to take a test and still do poorly, most likely anxiety is interfering with your ability to think.

For some very good assistance on how to start the process of learning, relearning, or studying mathematics in an encouraging and non-threatening way, read *Overcoming Math Anxiety* by Sheila Tobias or *Mind Over Math* by Stanley Kogelman and Joseph Warren. Both books tell how to study, learn, and solve word problems, fractions, statistics, percentages, algebra, and yes, even calculus.

# Appendix I

The course outlined below is taught at Mesa Community College in Mesa, Arizona. The five-week course begins about two weeks after the start of the semester so that it can be advertised in all beginning math classes. Math- and test-anxious students are encouraged to register for the course. It is taught on a credit/no-credit basis and does not count toward graduation.

## Course Prefix, Number, Title, Credits, and Periods

MA 066   *Math Anxiety and Avoidance*
One Credit, One Period

### Course Need

The course was developed for students who are afraid of or who avoid taking mathematics. Approximately 80% of college majors require at least one college-level math course; hence, math avoidance has serious limiting consequences. The course helps students reduce the anxiety which causes them to fail, withdraw from, or avoid mathematics.

## Course Description

The course is designed for students who avoid taking mathematics courses or who have anxiety in mathematics courses. Math anxiety is described, its etiology exposed, and anxiety reduction techniques are taught. How to study math as well as test-taking techniques are discussed. Positive math experiences are provided. Prerequisite: Concurrent enrollment in another math course.

## Course Competencies

The students will:

1. Write a paragraph describing the physical symptoms of their math anxiety.

2. Write a paragraph detailing how math anxiety disables them.

3. Describe in writing previous math experiences which have produced anxiety.

4. Write a hierarchy of math anxiety-producing stimuli to be used in desensitization.

5. Write a list of anxiety-producing self-talk messages.

6. Write a step-by-step prescription for studying mathematics.

7. Perform math activities while keeping anxiety at a minimum.

8. Complete all math assignments.

9. Complete a math final exam with a score of 70% or more.

# Course Outline

I. What is Math Anxiety?
   A. Description of the physiology of anxiety
   B. The relationship of anxiety to the cognitive processes
   C. How math anxiety is cognitively disabling

II. The Roots of Math Anxiety
   A. Math myths
   B. Former math experiences
   C. Self-defeating beliefs
   D. Self-fulfilling prophecies

III. Techniques for Reducing Math Anxiety
   A. The relation between cognition and emotion
   B. Self-talk exercise
   C. Relaxation and desensitization
   D. The role of classical conditioning in modifying object-related irrational fears
   E. The role of insight in behavior change

IV. Logical Consequences of Math Avoidance
   A. Career choices
   B. College majors

V. Study and Test-taking Techniques
   A. A prescription for studying mathematics
   B. The role of analysis, synthesis, and generalization in studying mathematics

VI. Measuring Change
   A. Suinn Math Anxiety Behavior Scale
   B. Revised Mathematics Attitude Scale

VII. Mathematics Topics
   A. Multiplication facts
   B. Fractions
      1. Common denominator in adding and subtracting
      2. Word problems using multiplication and division
   C. Percent problems
   D. Prime numbers and factors
   E. Arbitrary bases

VIII. Modular Arithmetic
   A. Designs based on modular arithmetic
   B. Logic and truth tables
   C. Metric system
   D. Computer literacy
   E. Hand calculators

# Appendix II

# RESOURCES

Allen, G.J. "Effectiveness of study counseling and desensitization in alleviating test anxiety in college students," *Journal of Abnormal Psychology*, 1971, 77, 282-289.

Allen, G.J. "Treatment of test anxiety by group administered relaxation and study counseling," *Behavior Therapy*, 1973, 4, 349-360.

Bandler, R. *Using Your Brain — for a Change*. Moab, Utah: Real People Press, 1985.

Bandura, A. *Principles of Behavior Modification*. New York: Holt, Rinehart, & Winston, 1969.

Bandura, A., & Barab, P. "Processes governing disinhibitory effects through symbolic modeling," *Journal of Abnormal Psychology*, 1973, 82, 1-9.

Bandura, A., Blanchard, E., & Ritter, B. "The relative efficacy of desensitization and modeling treatment approaches for inducing affective behavioral, and attitudinal changes," *Journal of Personality and Social Psychology*, 1969, 13, 173-199.

Bandura, A., Grusec, J., & Menlove, F. "Vicarious extinction of avoidance behavior," *Journal of Personality and Social Psychology*, 1967, 5, 16.

Beck, A. "Cognitive therapy: Nature and relation to behavior therapy", *Behavior Therapy*, 1970, 1, 184-200.

Beck, A. *Cognitive Therapy and Emotional Disorders*. New York: International Universities Press, 1976.

Bem, S. "Verbal self-control: The establishment of effective self-instruction," *Journal of Experimental Psychology*, 1967, 74, 485-491.

Blackwood, R. *Mediated Self-control: An Operant Model of Rational Behavior*. Akron, Ohio: Exordium Press, 1972.

Bloom, B., & Broder, L. *The Problem Solving Processes of College Students*. Chicago: University of Chicago Press, 1950.

Brown, B. "Cognitive aspects of Wolpe's behavior therapy," *American Journal of Psychiatry*, 1967, 124, 854-859.

Cautela, J. "Covert processes and behavior modification," *Journal of Nervous and Mental Disease*, 1973, 157, 27-35.

Claridge, G. *Personality and Arousal: A Psychophysiological Study of Psychiatric Disorder*. New York: Macmillan (Pergamon), 1967.

Clarek, D.E. "The treatment of monosymptomatic phobias by systematic desensitization," *Behavior Research and Therapy*, 1963, 1, 63.

Colter, S.B. "Sex differences and generalization of anxiety reduction with automated desensitization and minimal therapist interaction," *Behavior Research and Therapy*, 1970, 8, 273-285.

Davison, G. "Systematic desensitization as a counterconditioning process," *Journal of Abnormal Psychology*, 1968, 73, 91-99.

Davison, G., & Wilson, G. "Processes of fear-reduction in systematic desensitization: Cognitive and social reinforcement factors in humans," *Behavior Therapy*, 1973, 4, 1-21.

Dawsley, H.H., *et al.* "Self-administered desensitization on a psychiatric ward," *Journal of Behavior Therapy and Experimental Psychiatry*, 1973, 4, 301.

Deane, G. "Human heart rate responses during experimentally induced anxiety: A follow-up with controlled respiration," *Journal of Experimental Psychology*, 1964, 67, 193-195.

Deffenbacher, J.L. "Test anxiety: The problem and possible responses," *Canadian Counselor*, 1977, 11 (2), 59-64.

Denny, D.R., & Rupert, P.A. "Desensitization and self-control in the treatment of test anxiety," *Journal of Counseling Psychology*, 1977, 24, 272-280.

Donner, L. "Automated group desensitization: A follow-up report," *Behavior Research and Therapy*, 1970, 8, 241-247.

Donner, L., & Guerney, B.G. "Automated group desensitization for test anxiety," *Behavior Research and Therapy*, 1969, 7, 1-13.

Dweck, C. "The role of expectations and attributions in the alleviation of learned helplessness," *Journal of Personality and Social Psychology*, 1975, 31, 674-685.

D'Zurilla, T., & Goldfried, M. "Problem solving and behavior modification," *Journal of Abnormal Psychology*, 1971, 78, 107-126.

Ellis, A. *Reason and Emotion in Psychotherapy*. New York: Lyle Stuart Press, 1962.

Emery, J., & Krumholtz, J. "Standard versus individualized hierarchies in desensitization to reduce test anxiety," *Journal of Counseling Psychology*, 1967, 14, 204.

Epstein, S. "Natural healing processes of the mind," in H. Lowenheim (Ed.), *Meanings of Madness*. New York: Behavioral Publications, 1976.

Farber, I. "The things people say to themselves," *American Psychologist*, 1963, 18, 185-197.

Friedman, D.E. "A new technique for the systematic desensitization of phobic symptoms," *Behavior Research and Therapy*, 1966, 4, 139.

Gagne, R., & Smith, E. "A study of the effects of verbalization in problem solving," *Journal of Experimental Psychology*, 1962, 63, 12-18.

Garfield, Z., *et al*. "Effect of 'in vivo' training on experimental desensitization of a phobia," *Psychological Reports*, 1967, 20, 515.

Garlington, W.K., & Cotler, S.B. "Systematic desensitization of test anxiety," *Behavior Research and Therapy*, 1968, 6, 247-256.

Goldfried, M. "Systematic desensitization as training in self-control," *Journal of Consulting and Clinical Psychology*, 1971, 37, 228-234.

Goldfried, M. "Reduction of generalized anxiety through a variant of systematic desensitization," in M. Goldfried, & Merbaum (Eds.), *Behavior Change Through Self-control*. New York: Holt, Rinehart & Winston, 1973.

Goldfried, M., DeCanteceo, E., & Weinberg, L. "Systematic rational restructuring as a self-control technique," *Behavior Therapy*, 1974, 5, 247.

Goldfried, M., & Davison, G. *Clinical Behavior Therapy*. New York: Holt, Rinehart & Winston, 1976.

Goldfried, M., & Goldfried, A. "Cognitive change methods," in F. Kanfer, & A. Goldstein (Eds.), *Helping People Change*. New York: Pergamon Press, 1975.

Goldfried, M., & Sobocinski, D. "Effect of irrational beliefs on emotional arousal," *Journal of Consulting and Clinical Psychology*, 1975, 43, 504-510.

Goldfried, M., & Trier, C. "Effectiveness of relaxation as an active coping skill," *Journal of Abnormal Psychology*, 1974, 83, 348-355.

Hale, W., & Strickland, B. "Induction of mood states and their effects on cognitive and social behaviors," *Journal of Consulting and Clinical Psychology*, 1976, 44, 155.

Heckhausen, H. "Fear of failure as a self-reinforcing motive system," in I. Sarason, & C. Spielberger (Eds.), *Stress and Anxiety*, Volume II. Washington, D.C.: Hemisphere, 1975.

Henderson, A., Montgomery, I., & Williams, C. "Psychological immunization: A proposal for preventive psychiatry," *Lancet*, 1972, May, 1111-1112.

Horan, J., & Dellinger, J. "'In vivo' emotive imagery: A preliminary test," *Perceptual and Motor Skills*, 1974, 39, 359-362.

Inhelder, B., Sinclair, H., & Bovet, M. *Learning and the Development of Cognition.* Cambridge, Mass.: Harvard University Press, 1974.

Jacobson, E. *Progressive Relaxation.* Chicago: University of Chicago Press, 1938.

Janis, I. *Psychological Stress.* New York: John Wiley & Sons, 1958.

Janis, I. "Psychodynamic aspects of stress tolerance," in S. Klausner (Ed.), *The Quest for Self-control.* New York: Free Press, 1965.

Johnson, S., & Sechrest, C. "Comparison of desensitization and progressive relaxation in treating test anxiety," *Journal of Consulting and Clinical Psychology,* 1968, 32, 280-286.

Kagan, J. "Reflection-impulsivity: The generality and dynamics of conceptual tempo," *Journal of Abnormal Psychology,* 1966, 71, 17-24.

Kahn, M., & Baker, B. "Desensitization with minimal therapist contact," *Journal of Abnormal Psychology,* 1968, 73, 198.

Katahm, M., *et al.* "Group counseling and behavior therapy with test anxious college students," *Journal of Consulting Psychology,* 1966, 30, 544.

Kazdin, A. "Covert modeling and the reduction of avoidance behavior," *Journal of Abnormal Psychology,* 1973, 81, 87-95.

Kogelman, S. & Warren, J. *Mind Over Math.* New York: Dial Press, 1978.

Lang, P. "Fear reduction and fear behavior: Problems in treating a construct," in J. Shlien (Ed.), *Research in Psychotherapy,* Volume 3. Washington, D.C.: APA, 1968.

Lang, P. "The mechanics of desensitization and the laboratory study of human fear," in C. Franks (Ed.), *Assessment and Status of the Behavior Therapies.* New York: McGraw-Hill, 1969.

Lang, P.J., & Lazovik, A.D. "The experimental desensitization of a phobia," *Journal of Abnormal Psychology,* 1963, 66, 519-525.

Lang, P., Lazovik, A., & Reynolds, D. "Desensitization, suggestibility, and pseudotherapy," *Journal of Abnormal Psychology,* 1965, 70, 395-402.

Lazarus, A.A. "Group therapy of phobic disorders by systematic desensitization," *Journal of Abnormal and Social Psychology,* 1961, 63, 504.

Lazarus, R., & Alfert, E. "Short-circuiting of threat by experimentally altering cognitive appraisal," *Journal of Abnormal and Social Psychology,* 1964, 69, 195-205.

Lent, R.W., & Russell, R.K. "Treatment of test anxiety by cue-controlled desensitization and study-skills training," *Journal of Counseling Psychology,* 1978, 25, 217-224.

London, P. "The end of ideology in behavior modification," *American Psychologist,* 1972, 27, 913-920.

Mahoney, M. *Cognition and Behavior Modification.* Cambridge, Mass.: Ballinger Publishing Co., 1974.

McKeachie, W. "The decline and fall of the laws of learning," *Educational Researcher*, 1974, 3, 7-11.

Meichenbaum, D. "Examination of model characteristics in reducing avoidance behavior," *Journal of Personality and Social Psychology*, 1971, 17, 298-307 (b).

Meichenbaum, D. "Cognitive modification of test anxious college students," *Journal of Consulting and Clinical Psychology*, 1972, 39, 370-380.

Meichenbaum, D. *Cognitive Behavior Modification.* New York: Plenum Press, 1977.

Meichenbaum, D. "Self instructional approach to stress management," in C.D. Spielberger, & I.G. Sarason (Eds.), *Stress and Anxiety.* Washington, D.C.: Hemisphere Publishing Company, 1975.

Meichenbaum, D. "Toward a cognitive theory of self-control," in G. Schwartz, & D. Shapiro (Eds.), *Consciousness and Self Regulation*, Volume 1. New York: Plenum Press, 1976 (b).

Meichenbaum, D., & Cameron, R. "The clinical potential of modifying what clients say to themselves," *Psychotherapy: Theory, Research, and Practice*, 1974, 11, 103-117.

Meichenbaum, D., & Turk, D. "The cognitive-behavioral management of anxiety, anger and pain," in P. Davidson (Ed.), *The Behavioral Management of Anxiety, Depression and Pain.* New York: Bruner Mazel, 1976.

Melnick, J., & Russell, R.W. "Hypnosis versus systematic desensitization in the treatment of test anxiety," *Journal of Counseling Psychology*, 1976, 23, 291-295.

Mitchell, C. "Improving math performance by reducing test anxiety," *Arizona Personnel and Guidance Journal*, 1982, 8, 19-23.

Mitchell, K.R., & Ng, K.T. "Effects of group counseling and therapy on the academic achievement of test-anxious students," *Journal of Counseling Psychology*, 1972, 19, 491-497.

Morris, L., & Liebert, R. "Relationship of cognitive and emotional components of test anxiety to physiological arousal and academic performance," *Journal of Consulting Clinical Psychology*, 1970, 35, 332-337.

Mowrer, O., & Viek, P. "An experimental analogue of fear from sense of helplessness," *Journal of Abnormal and Social Psychology*, 1948, 43, 193-200.

Osterhouse, R.A. "Desensitization and study skills: Two types of test-anxious students," *Journal of Counseling Psychology*, 1972, 19, 301-307.

Paul, G.L. *Insight Versus Desensitization in Psychotherapy.* Stanford: Stanford University Press, 1966.

Paul, G. "Outcome of systematic desensitization II: Controlled investigations of individual treatments, techniques, variations, and current status," in C. Franks (Ed.), *Behavior Therapy: Appraisal and Status.* New York: McGraw-Hill, 1969.

Paul, G., & Shannon, D.T. "Treatment of anxiety through systematic desensitization in therapy groups," *Journal of Abnormal Psychology*, 1966, 71, 121-135.

Paul, G., and Erikson, S. "Effects of test anxiety on 'real-life' examinations," *Journal of Personality*, 1964, 32, 480.

Pervin, L. "The need to predict and control under conditions of threat," *Journal of Personality*, 1963, 31, 570-585.

Phillips, G., *et al.* "Self-administered systematic desensitization. *Behavior Research and Therapy*, 1972, 10, 93.

Plutchik, R., and Ax, A. "A critique of 'Determinants of emotional states' by Schachter and Singer (1962)," *Psychophysiology*, 1967, 4, 79-82.

Rachman, S. "Systematic desensitization," *Psychological Bulletin*, 1967, 67, 93-103.

Register, B.W., Stockton, R.A., & Maultsby, M.C. "Counseling the test anxious: An alternative," *Journal of College Student Personnel*, 1977, 18, 506-510.

Repucci, N., & Baker, B. "Self desensitization: Implications for treatment and teaching," in R.D. Rubin, & C.M. Frank (Eds.). New York: Academic Press, 1969.

Richardson, F.C., & Suinn, R.M. "Effects of two short-term desensitization methods in the treatment of test anxiety," *Journal of Counseling Psychology*, 1974, 21, 457-458.

Rimm, D., & Litvak, S. "Self-verbalization and emotional arousal," *Journal of Abnormal Psychology*, 1969, 74, 181-187.

Rosen, G., *et al.* "A controlled study to assess the clinical efficacy of total self-administered systematic desensitization," *Journal of Consulting and Clinical Psychology*, 1976, 44, 208.

Russell, R.K., & Sipich, J.F. "Cue-controlled relaxation in the treatment of test anxiety," *Journal of Behavior Therapy and Experimental Psychiatry*, 1973, 4, 47-49.

Sarason, I. "Test anxiety and cognitive modeling," *Journal of Personality and Social Psychology*, 1973, 28, 58-61.

Schachter, S. "The interaction of cognitive and physiological determinants of emotional state," in C. Spielberger (Ed.), *Anxiety and Behavior.* New York: Academic Press, 1966.

Schwartz, G. "Cardiac responses to self-induced thoughts," *Psychophysiology*, 1971, 8, 462-467.

Sherrington, C. *Integrative Action of the Nervous System*. New Haven: Yale University Press, 1980.

Sipich, J.F., & Deffenbacher, J.L. "Comparison of relaxation as self-control and systematic desensitization in the treatment of test anxiety," *Journal of Consulting and Clinical Psychology*, 1977, 45, 1202-1203.

Skaggs, E. "Changes in pulse, breaking and steadiness under conditions of startledness and excited expectancy," *Journal of Comparative and Physiological Psychology*, 1926, 6, 303-318.

Skinner, B.F. *Science and Human Behavior*. New York: Macmillan, 1953.

Solyom, L., & Miller, S. "Reciprocal inhibition by aversion relief in the treatment of phobias," *Behavior Research and Therapy*, 1967, 5, 313-324.

Spiegler, M., *et al.* "Cognitive and emotional components of test anxiety: Temporal factors," *Psychological Reports*, 1968, 22, 451.

Spiegler, M., Cooley, E., Marshall, G., Prince, H., Puckett, S., & Slenazy, J. "A self-control versus a counterconditioning paradigm for systematic desensitization: An experimental comparison," *Journal of Counseling Psychology*, 1976, 23, 83-86.

Spielberger, C. "The effects of manifest anxiety on the academic achievement of college students," *Mental Hygiene*, 1962, 46, 420.

Stewart, M.A. "Psychotherapy by reciprocal inhibition," *American Journal of Psychiatry*, 1961, 188, 175.

Suinn, R.M., Edie, C., Nicoletti, J., & Spinelli, R.R. "Automated short-term desensitization," *Journal of College Student Personnel*, 1973, November, 471-476.

Suinn, R., & Richardson, F. "Anxiety management training: A nonspecific behavior therapy program for anxiety control," *Behavior Therapy*, 1971, 2, 498-510.

Tobias, S. *Overcoming Math Anxiety*. New York: Norton, 1978.

Valins, S., & Ray, A. "Effects of cognitive desensitization on avoidance behavior," *Journal of Personality and Social Psychology*, 1967, 7, 345-350.

Watson, J.B., Raynor, R. "Conditioned emotional reactions," *Journal of Experimental Psychology*, 1920, 3, 1-14.

Wilkins, W. "Desensitization: Social and cognitive factors underlying the effectiveness of Wolpe's procedure," *Psychological Bulletin*, 1971, 76, 311-317.

Wolberg, L. *Medical Hypnosis*. New York: Grune & Stratton, 1948.

Wolpe, J., Brady, J.P., Serber, M., Agras, W.S., & Liberman, R.P. "The current status of systematic desensitization," *American Journal of Psychiatry*, 1973, 130, 961-965.

Wolpe, J. *The Practice of Behavior Therapy*. Elmsford: Pergamon Press Inc., 1969.

Wolpe, J. "Reciprocal inhibition as the main basis of psychotherapeutic effects," *Arch. Neurological Psychiatry*, 1954, 72, 205.

Wolpe, J. *Psychotherapy by Reciprocal Inhibition.* Stanford: Stanford University Press, 1958.

Wolpe, J. "The systematic desensitization treatment of neuroses," *Journal of Nervous and Mental Disease*, 1961, 112, 189.

Wolpe, J. "Quantitative relationships in the systematic desensitization of phobias," *American Journal of Psychiatry*, 1963, 119, 1062.

Wolpe, J., & Lazarus, A. *Behavior Therapy Techniques: A Guide to the Treatment of Neuroses.* London: Pergamon Press, 1966.

Wolpin, M., & Raines, J. "Visual imagery, expected roles and extinction as possible factors in reducing fear and avoidance behavior," *Behavior Research and Therapy*, 1966, 4, 25.

Wood, D., & Obrist, P. "Effects of controlled and uncontrolled respiration on the conditioned heart rate in humans," *Journal of Experimental Psychology*, 1964, 69, 221-229.

Yates, D. "Relaxation in psychotherapy," *Journal of General Psychology*, 1946, 34, 213-238.

Zemore, R. "Systematic desensitization as a method of teaching a general anxiety-reducing skill," *Journal of Consulting and Clinical Psychology*, 1975, 43, 157.

# About the Authors

Charlie Mitchell is a psychologist in private practice in Arizona. He was a counselor and teacher at Mesa Community College in Mesa, Arizona for 19 years. One of the courses he developed there was Math Anxiety and Avoidance. Dr. Mitchell has a Ph.D. degree in Counseling from the University of Arizona, and B.S. and M.A. degrees in psychology from Eastern New Mexico University. He is also the author of *Career Exploration: A Self-paced Approach* and *Job Hunting: A Self-directed Guide.*

Lauren Collins has been a classroom teacher, a bilingual teacher aide trainer, an editor, an author, a community college instructor, and currently a child and family therapist. She has a B.A. degree from Smith College, and an M.Ed. from Antioch University. She is also the author of *Job Hunting: A Self-directed Guide* and *La participación de los padres en la educación de sus hijos.*